12.00
1st

EARTH AND YOU

EARTH AND YOU
Tales of the Environment

Charles Officer and Jake Page

Peter E. Randall Publisher
Portsmouth, New Hampshire
2000

© 2000 by Charles Officer and Jake Page
Printed in the United States of America

Peter E. Randall Publisher
Box 4726, Portsmouth, NH 03802-4726

Distributed by University Press of New England
    Hanover and London

Designed by Sarah C. Foley

Library of Congress Cataloging-in-Publication Data

Officer, Charles B.
  Earth and you : tales of the environment / Charles Officer and Jake Page.
    p. cm.
  Includes bibliographical references and index.
  ISBN 0-914339-87-7
  1. Environmental degradation. 2. Environmental protection. I. Page, Jake. II. Title.
  GE140 .O34 2000
  363.7--dc21

                                                00-26347

CONTENTS

PREFACE vii

PART I—Yesterday and Today

CHAPTER 1
We, as a Species, have Grown to Dominate the Earth—to Alter its Landscape and to Eliminate Natural Wilderness Areas
    Introduction 1
    National Parks, Forest Preserves and Wildlife
        Refuges—our national heritage 1
    Haiti—deforestation 16
    Brazil—tropical rain forests 21

CHAPTER 2
...To Overextend the Limits of Water and Land Use,
    Venice—disappearance of a venerable city 31
    Aral Sea—requiem for an inland sea 41
    Dust Bowl—desertification 47
    California and Colorado River—water depletion 56

CHAPTER 3
...And To Decimate or Drive to Extinction Other Species.
    Buffaloes and marine mammals—animal overkill ... 63
    Passenger pigeon and dodo—bird overkill 70
    Gypsy moth and zebra mussel—introduced species . 75

CHAPTER 4
But There are Hopeful Signs, and Corrective Actions have been Taken, Largely through the Efforts of Individuals—On a Local Scale,
    Love Canal—Lois Gibbs 83
    Times Beach—Judy Piatt 97
    Cuyahoga River—Ben Stefanski 99
    Centralia—Joan Girolami 102

CHAPTER 5
...On a Regional Scale
    London and Los Angeles—Harold Des Voeux 113
    Lake Erie and Chesapeake Bay—
        Chesapeake Bay Foundation 122

    Acid rain—Noye Johnson 131
    Ogallala Aquifer—Groundwater
        Management Districts 135

CHAPTER 6
...And Even on a National to Global Scale.
    Pesticides—Rachel Carson 141
    Ozone layer depletion—Mario Molina
        and Sherwood Rowland 151
    Global warming, part I—Roger Revelle 160
    Global warming, part II—Wallace Broecker 171
    Green revolution—Norman Borlaug 183

PART II—Tomorrow

CHAPTER 7
There is Much More to be Done and it Will Require the Concerted Efforts of All of Us.
    Introduction 195
    Past successes and failures 195
    Global environmental problems 200

CHAPTER 8
There are two Major Environmental Problems that have to be Addressed—Global Population, A Sociological Problem,
    Malthus Theorem 203
    History of population growth 205
    Birth control, female education 210
    Moral and religious leadership 212
    Family Planning 217
    International initiative 218

CHAPTER 9
...And Alternative Energy Sources, a Scientific Problem.
    Natural resources depletion 221
    Research and development 226
    Alternative energy sources 229
    National initiative 231

REFERENCES 237

INDEX 251

PREFACE

What the world will be like in fifty years is anybody's guess. The most successful predictions about the future state of the globe or human affairs are usually the vaguest, like the prophecies of Nostradamus and other mystics whose words can be interpreted as one wishes and bent to fit almost any future that arises.

Other predictions are, perhaps, more reliable—those that emerge from the contemplation of cause and effect. For example, that an industrializing world dependent chiefly upon oil, gas and coal will one day run out of these materials is an unavoidable prediction. Such predictions can be, and are usually intended to be, curative—they can, in fact, change the course of the future if they are heeded.

This book is largely optimistic. It tells the stories of how a number of key environmental problems, here and abroad, have been recognized and dealt with, often with success, often at the instigation of single individuals from visionary U.S. presidents to outraged mothers. Human affairs, after all—both problems and solutions—do not come about of their own accord. On local, regional, national and even international levels, environmental problems in great variety have been successfully addressed by determined people and, quite often, solved.

In most cases, however, the solutions are temporary—like mowing dandelions instead of uprooting them—because most of the planet's environmental problems are ultimately driven by two huge engines. These are 1) population growth of an unprecedented and unsustainable magnitude and 2) the unchecked exhaustion of the Earth's resources, particularly energy resources that are limited in supply, and their conversion into substances that are cumulatively toxic—to human society, to life itself, and even the planet's climate.

To ignore these large matters (we are all so busy with our own lives, our own daily challenges) would be criminal. And if they do go ignored, the results are predictable in outline at least: starvation, illness, loss, degradation, effects of global warming, tragedy of many kinds on a scale that is barely imaginable.

So this book is yet another voice calling for our best efforts. We are optimistic that they will be forthcoming. We have to be optimistic: consider the alternative.

PART I
Yesterday and Today

CHAPTER 1

We, as a Species, have Grown to Dominate the Earth—to Alter its Landscape and to Eliminate Natural Wilderness Areas.

Introduction

Let us start with the past. What has been our history as a country and as a member of the global community in addressing environmental concerns? There have been environmental degradations but there have also been much conservation and preservation and corrective actions to environmental problems. And these positive aspects have largely come about through the actions of individuals. Some have been scientists and engineers but many have been individuals who were concerned about a particular problem or, inadvertently, became a pawn in an environmental mishap. All in all—in our opinion—the record is good. We have become an environmentally concerned public.

National Parks, Forest Preserves and Wildlife Refuges—Our National Heritage

It is always nice to start out with a success story. And that is the case for the National Parks, the National Forests, and the Fish and

Wildlife Refuges. They are there because of the farsighted and dedicated work of a few individuals who foresaw the national need. They are part of our preservation and conservation heritage and the envy of all other industrialized nations.

The agency principally responsible for the National Forests is the Forest Service of the Department of Agriculture. That for the National Parks, the National Park Service and that for the Fish and Wildlife Refuges, the Fish and Wildlife Service, both of the Department of Interior. All three are superlative agencies and are run in Washington and in the field by dedicated personnel. The total areas managed by these three agencies are given in the accompanying table. The total area of 353,000,000 acres is enormous; it amounts to 18% of the total land and water area of the continental United States. And this does not include the state, regional and local parks, forests and wildlife refuges (see Table 1).

All of this came about through government action stimulated by individuals beginning in the middle to late 1800s, gaining impetus in the early to middle 1900s, and continuing on to today. Of the many individuals who have made significant contributions through the years we mention, here, three who were particularly instrumental in the early days in fostering the concepts of the National Parks, National Forests, and Fish and Wildlife Refuges—respectively, John Muir, Gifford Pinchot, and Jay Northwood Darling. We will return later to their specific efforts.

The jurisdiction of the National Park Service has grown over the years to include not only the National Parks but a whole slew of other entities. As of 1988, their responsibility included that for the 50 National Parks such as Yosemite, Yellowstone and the Great

Agency	Acres
Fish and Wildlife Service	91,000,000.
Forest Service	189,000,000.
National Park Service	<u>73,000,000.</u>
Total	353,000,000.

From World Almanac, 1993.

Table 1. Public lands administered by federal agencies

Smoky Mountains; 29 National Historical Parks such as the Chesapeake and Ohio Canal and Valley Forge; 24 National Battlefields, Battlefield Parks, Battlefield Sites and Military Parks such as Gettysburg, Shiloh and Cowpens; 24 National Memorials such as the Lincoln Memorial and Mount Rushmore; 69 National and International Historic Sites such as the Carl Sandburg Home and Saint-Gaudens Home; 80 National Monuments such as the Statue of Liberty and Canyon de Chelly; and 56 National Preserves, Seashores, Parkways, Lakeshores, Rivers, and Scenic Rivers and Riverways such as Cape Cod, Natchez Trace and Wolf Trap Farm for the Performing Arts.

The total number of visitors to all areas administered by the National Park Service amounted to 283,000,000 people in 1988. For comparison, the total population of the United States in 1988 was 245,000,000. In other words, more people visit these National Park and associated areas in one year than live in the United States, quite a credit to the Park Service and to us.

The first national park, in fact but not in name at the time when it was first set aside, is Yosemite. It was established by an act of Congress and signed into law by President Lincoln on 30 June 1864, a date during the Civil War when Sherman was on his march through Georgia and Lee was under siege at Petersburg. In the middle 1800s little of the scenic beauty of the West and particularly that of Yosemite were known to those who lived in the East. California was known for its gold rush and little else. That deficiency was corrected by several individuals. Among them was Horace Greeley, the famous editor, who visited Yosemite Valley in 1859 and declared it "the most unique and majestic of nature's marvels". Another was Albert Bierstadt, the equally famous landscape artist, who painted an impressive picture of the Valley in 1863.

The preservation of Yosemite Valley and its associated grove of Sierra redwoods became a concern of a small group of Californians including Israel Ward Raymond. On 20 February 1864 Raymond wrote to California's United States Senator John Conness that Yosemite should be preserved "for public use, resort and recreation" and that these actions should be taken "inalienable forever". Conness argued persuasively before Congress, resulting in the Yosemite Act of 1864. At that time Congress turned Yosemite over to the state of California for administration.

In the Yosemite Act of 1890 these lands were extended for the protection of the giant sequoias into what have become Kings Canyon and Sequoia National Parks. In 1905 California ceded back to the federal government the Yosemite lands.

Although Yosemite is now inundated every year by tourists to such an extent that its visitor areas resemble an urban sprawl, it still remains one of the scenic wonders of the West. The Valley, itself, was formed by mountain glaciation in the Sierras during the recent (geologic) Ice Ages over the past 300,000 years or so when the Laurentide ice sheets extended across Canada and the northern United States. There are at least three periods of mountain glaciation at Yosemite followed by warmer interglacial periods. (We are presently in one of these warmer interglacial periods.) As the glaciers moved slowly down the mountain slopes, they carved out U-shaped valleys, one of these being Yosemite

Bridal Veil Falls, Yosemite National Park. From Harris, 1975. United States Geological Survey, Denver.

Valley, another being the adjacent Hetch Hetchy Valley. There were also smaller, tributary glaciers to the main glacier, much the same as we have with our present day river systems. The main, Yosemite glacier gouged deeper than the tributary glaciers that occupied the side valleys. When the last glaciers retreated and melted, the floors of the tributary glaciers were left at a considerable elevation above the floor of Yosemite Valley, leading to the hanging valleys and spectacular waterfalls that are now part of Yosemite National Park. Yosemite actually has over half of the highest waterfalls in the United States. For example, Bridal Veil Falls, one of the more picturesque, has a drop of 620 feet as compared with 190 feet for Niagara Falls.

The first national park, as such, is Yellowstone. It was established by an act of Congress and signed into law by President Grant in 1872. Yellowstone marks the beginning of the national park system. Yellowstone is huge, 2,220,000 acres, as compared with Yosemite, 760,000 acres. It is the largest National Park within the contiguous forty-eight states. (There are several larger National Parks in Alaska.)

The wonders of the geysers, waterfalls and canyons of the Yellowstone area in the Rocky Mountains were brought to the attention of those who lived in the East as a result of two expeditions. One, in 1870, known as the Washburn-Langford-Doane expedition; and the other, in 1871, known as the Hayden expedition. Henry Washburn was, at that time, surveyor-general of the Montana territory; Nathaniel P. Langford, an influential citizen of Montana; and Gustavus Doane, a second lieutenant in the Army and leader of the military escort party. Ferdinand Hayden was a professor of geology at the University of Pennsylvania and director of the United States Geological and Geographical Survey of the Territories, the predecessor agency to the present United States Geological Survey.

Doane's report received widespread publicity in the newspapers and magazines and Langford lectured extensively in the East following their expedition. Langford proclaimed that "a grander scene than the lower cataract of the Yellowstone was never witnessed by mortal eye. It is a sheer, compact, solid, perpendicular sheet, faultless in all the elements of grandeur and picturesque beauties" and further that the Old Faithful geyser was the "new and, perhaps, most remarkable feature in our scenery and physical history". For his efforts in promoting the preservation of Yellowstone as a national heritage Langford earned the sobriquet "National Park Langford". He served as superintendent of the park from 1872 to 1877 without pay.

The Hayden expedition was scientific in nature including individuals knowledgeable in topography, meteorology, mineralogy, and zoology. His report and appearances before Congress were an important part in establishing Yellowstone as the first National Park. He argued persuasively that Yellowstone should not go the way of Niagara Falls "to fence in these rare wonders so as to charge visitors a fee, as is now done at Niagara Falls, for the sight of that which ought to be as free as the air or water". The Hayden expedition also included the artist, Thomas Moran. His panoramic painting, seven by twelve feet, of the Grand Canyon of the Yellowstone was purchased by Congress in 1872 and later hung in the Senate lobby.

One of those influential in setting aside Yellowstone as a National Park was the financier, Jay Cooke. He paid the expenses for Thomas Moran to accompany the Hayden expedition. He is attributed, by some, for suggesting the park bill and was influential in its passage. The year 1872 may have been a good year for Cooke with the passage of the Yellowstone Act; the year 1873 certainly was not. His banking company had invested heavily in the construction of the Northern Pacific Railway and was overextended. His failure and that of several other eastern banks led to Black Friday and the financial panic of 1873. His company was compelled to go into bankruptcy. He later repaid his creditors and recouped his fortune in western silver mines.

Yellowstone is best known for its geysers, such as Old Faithful, and its hot springs, such as Mammoth Hot Springs. A hot spring is simply that, a bubbling up through a vertical conduit of hot groundwater of 100°F or higher which has been heated at depth by a magma chamber of molten igneous material. A geyser is an intermittent hot spring. In this case there are side chambers to the main conduit. Superheated water and steam are trapped in the side chambers. At a critical point in the heating process a slight change in temperature or pressure will trigger a water and steam eruption. The system will then return to its relatively cooler state and the process will start again.

Much of Yellowstone is of volcanic origin and it is underlain by a giant caldera about 40 miles in diameter. A caldera is a basin-shaped depression or collapse structure left after a large volcanic eruption. The last major eruption at Yellowstone, and the last major eruption in the continental United States, was 600,000 years ago. It spread an ash layer of an inch or so in thickness over most of the United States west of the Mississippi and may have wiped out most of the vegetation in this vast region. The existence of geysers and hot springs and the current ground movement and swarms of minor earthquakes at Yellowstone suggest that it could happen again at some point in the future.

Following the establishment of Yellowstone as the first National Park there was a hiatus of new parks for the next twenty-seven years with the exception of the lands added to Yosemite. Over the following twenty year period, however, from 1899 to 1918 there was a spate of new parks, a bulwark of scenic preservation in our present day park system. They include Mount

Grand Canyon of the Yellowstone, Yellowstone National Park. From Harris, 1975. United States Geological Survey, Denver.

Rainier in 1899; Crater Lake, 1902; Mesa Verde, 1906; Grand Canyon, 1908; Glacier, 1910; Rocky Mountain, 1915; Hawaii, 1916; Lassen Volcanic, 1916; Acadia, 1916; Mount McKinley, 1917; Zion, 1918; and Katmai, 1918. All of these except for Mount Rainier were established during the administration of Theodore Roosevelt, 1901-1909, our most conservation-minded president, and his successors, Taft, 1909-1913, and Wilson, 1913-1921.

Another aspect of preservation came about with the passage of the Preservation of American Antiquities Act in 1906. As people began to settle in the Southwest, an unfortunate story of vandalism began to develop, specifically the looting of archaeological ruins of past Indian cultures. Jesse Nusbaum, who was for many years superintendent of Mesa Verde National Park, described the conditions succinctly. "The heyday, during the 1890s, of wholesale commercial looting of archaeological sites in the Southwest by 'pot hunters' to meet increasing market demands for artifacts and comprehensive collections caused prodigious damaging, destruction and loss of archaeological sites and values, since these pot hunters sporadically searched ruin sites solely for maximum salable loot by the most expeditious methods of unscientific excavation." Among those disturbed was John Lacey, an Iowa congressman and staunch preservationist. He introduced and pushed through to passage the Antiquities Act.

The act called for preservation of "objects of historic or scientific interest that are situated upon the lands owned or controlled by the Government of the United States". Roosevelt, as president, interpreted this clause broadly to include areas of geologic (scientific) as well as man-made significance. He immediately set aside in 1906 such features as Devil's Tower, Wyoming; Petrified Forest, Arizona; and Inscription Rock, New Mexico, as the first of our now several National Monuments.

One of the leading proponents of preservation in the early 1900s, when the movement achieved much, was John Muir. Muir was an explorer, naturalist, and writer. Born in Scotland to a strict Presbyterian family, he grew up in Wisconsin on a farm. He received little formal education but was a natural inventor of mechanical devices. He spent his early years working in machine shops in Wisconsin, Ontario, and Indiana. But he was not satisfied with this sort of life; he had an inner restlessness—a wanderlust. He expressed his confusion and dilemma in a letter to a friend at that time as follows.

> I *feel* something within, some restless fires that urge me on in a way very different from my real wishes, and I suppose that I am doomed to live in some of these noisy commercial centers.
>
> Circumstances over which I have no control almost compel me to abandon the profession of my choice, and to take up the business of an inventor, and now that I am among machines I *feel* that I have some talent that way, and so I almost think, unless things change soon, I shall turn my whole mind into that channel.

Finally, the wanderlust won out and in 1867 at the age of twenty-nine he began a trek which would eventually lead him to San Francisco and Yosemite. It started with a 1,000 mile walk from Indianapolis, Indiana, to Cedar Key, Florida. His naturalist as well as his writing instincts were aroused on the long walk. In the journal that he kept he expressed his wonders of the Cumberlands in a style which would later make him famous:

> Such an ocean of wooded, waving, swelling mountain beauty and grandeur is not to be described. Countless forest-clad hills, side by side in rows and groups, seemed to be enjoying the rich sunshine and remaining motionless only because they were so eagerly absorbing it. All were united by curves and slopes of inimitable softness and beauty.
>
> Oh, these forest gardens of our Father! What perfection, what divinity, in their architecture! What simplicity and mysterious complexity of detail! Who shall read the teachings of these sylvan pages, the glad brotherhood of rills that sing in the valleys, and all the happy creatures that dwell in them under the tender keeping of a Father's care?

…and further on the lowly creatures of the Earth:

> How narrow we selfish, conceited creatures are in our sympathies! How blind to the rights of all the rest of creation! With what dismal irreverence we speak of our fellow mortals! Though alligators, snakes, etc., naturally repel us, they are not mysterious evils. They dwell happily in these flowery wilds, are part of God's family, unfallen, undepraved, and cared for with the same species of tenderness and love as is bestowed on angels in heaven or saints on earth.

From Cedar Key he traveled by ship to Cuba, then to New York, then to Panama, across the Panama isthmus by train, and on to San Francisco. He finally settled in the San Francisco area with numerous trips to his favorite place of solace, Yosemite, as well as explorations to Alaska and the Southwest.

He spent much of his first few years in California at Yosemite working as need be as an itinerant laborer. When he finally left the valley five years later, he was a naturalist and scientist of standing, the acknowledged expert on life in the Sierras, and a writer of reputation. Louis Agassiz, the renowned glaciologist, wrote to Muir that he was the first man that Agassiz had heard of who had an adequate conception of glaciers. Ralph Waldo Emerson camped with him at Yosemite in 1871 as did Theodore Roosevelt in 1903. He wrote extensively for the San Francisco, New York, and Boston newspapers and the national magazines on the wonders of Yosemite and the Sierras and on environmental protection in general.

He was a mountaineer and a chronicler of what he saw. More than anyone else at that time he portrayed to the public at large the grandeur of the West and the importance of preserving it for future generations while the chance still remained. His own words best summarize his continuing and unassuming attitude as his fame increased and the accolades began to multiply.

> As long as I live, I'll hear waterfalls and birds and wind sing. I'll interpret the rocks, learn the language of flood, storm, and the avalanche. I'll acquaint myself with the glaciers and wild gardens and get as near to the heart of the world as I can.

Not all of Muir's preservationist efforts were successful and not all parks remain "inalienable forever". Hetch Hetchy Valley, located about 15 miles north of Yosemite Valley and within the Yosemite lands, is a case in point. It is similar in morphology to Yosemite Valley, a glacial erosion feature. San Francisco has always been in need of municipal water. In 1901 the mayor of San Francisco petitioned the Interior Department to build a dam across the valley as a water supply for the city and for hydroelectric power. The petition was denied at that time but renewed again in 1905, granted in 1908 with congressional and presidential approval in 1913.

During the deliberations the dam had the support of Gifford Pinchot, an ardent conservationist but not a preservationist, and the opposition of the Sierra Club. The Sierra Club had been

Teddy Roosevelt and John Muir in Yosemite, May, 1903. From Turner, 1985. Bancroft Library, University of California, Berkeley.

formed by Muir and a small group of like-minded people in 1892. (The Sierra Club has since grown to a membership of over 600,000 and is one of the more active advocate groups for environmental protection in the country.) It is somewhat of a paradox that Hetch Hetchy lake (reservoir) is a serene pastoral area for part of the year as compared with the overcrowded conditions at Yosemite Valley. Unfortunately for the rest of the year during periods of low water level, it is a mudflat lined with tree stumps.

Although Pinchot was not a preservationist, he and Theodore Roosevelt were the prime movers in conserving the nations forests and establishing, the National Forest system. Pinchot's background and early life were about as far removed from Muir's as one can find. He came from a wealthy family, was part of the Eastern establishment, and graduated from Yale. He summarized his philosophy in the following words.

> My own money came from unearned increment on the land in New York held by my grandfather, who willed the money, not the land to me. Having got my wages in advance in that way, I am now trying to work them out.

He went on to have two successful careers, one in forestry and the other in politics. Following his graduation from Yale in 1889, he studied forestry in Europe and became America's first trained forester. For many years during the 1890s and early 1900s he held the position of Forester and was director of the Division of Forestry in the Department of Agriculture. He was outspoken and reluctant to compromise, a bit of a curmudgeon. Few spoke of him in neutral terms; he was either admired or despised. Following his term as Forester he went on to run twice, unsuccessfully, for the United States Senate from Pennsylvania and twice, successfully, for governor of that state.

He was violently opposed to the practice of "chop and run", which completely leveled vast areas of virgin timber and left nothing but devastation behind. He considered forests as an agricultural crop. His view was that a forest could be both harvested and conserved by selective cutting, leaving seedlings and larger trees at regularly spaced intervals to provide seeds for future crops. He further considered that such practices could only be assured by government control of private cutting.

One of Pinchot's most important attributes was that he was a close friend and advisor to President Roosevelt—like-minded individuals on the subject of forest conservation. In a short period of time these two individuals established our present National Forests. In 1898, when Pinchot entered government service, there were 19 National Forests with 20,000,000 acres. In 1909, when Roosevelt left office, there were 149 National Forests with 193,000,000 acres, slightly larger than the present lands administered by the Forest Service.

The last of our conservation enterprises, the Fish and Wildlife Refuges, was also the result of a few dedicated individuals—notable among them being Jay Northwood Darling. Darling was a nationally syndicated and Pulitzer prize winning political cartoonist. He is better known by the abbreviation of his last name, Ding, with which he signed his cartoons. He was a midwesterner and his career extended from the early to the middle 1900s.

He was also a conservationist. In 1934 he was asked by President Franklin Roosevelt to be a member of a special presidential committee of three to study and recommend a program for the restoration and conservation of migratory waterfowl. Following that Roosevelt asked Darling to take over the position as head of the Bureau of the Biological Survey, the forerunner agency to the present Fish and Wildlife Service. (Darling was a conservative Republican from Iowa, had been a delegate to the Republican National Convention in 1932, and had campaigned for his friend, Herbert Hoover and Roosevelt's opponent, in the following general election.)

Darling's stay with the Biological Survey was short, only two years; but he accomplished much in rejuvenating it and accelerated the country's actions towards acquiring and setting aside lands as wildlife refuges. He acted in a forthright and demanding manner; he was not part of the political or civil service system and was not dependent on it.

His iconoclastic approach can best be illustrated by two examples. Darling had put together his funding needs for acquiring wildlife refuges. He had obtained the approval of the Secretary of

Agriculture, Henry Wallace, and the Secretary of the Interior, Harold Ickes, but was informed that he would also need the approval of Harry Hopkins, special assistant to the President. After his staff's presentation to Hopkins, Hopkins responded in a blasé manner, "I don't know if we're interested in the relief of birds." At that Darling jumped to his feet and shouted at Hopkins, "Harry, I was a trustee at Grinnell College when you were just a student!" and continued with some further words that Hopkins had strongly encouraged him to accept the position as head of the Biological Survey. Hopkins replied, "Where do I sign these papers?" and then turning to his assistants commented, "See, that's how things get done in Washington."

In a milder exchange with the president he petitioned for the reinstatement of a $4,000,000 additional appropriation for the Biological Survey. He wrote, with regard to retired agricultural land,

> Others just grow grass and trees on it. We grow grass, trees, marshes, lakes, ducks, geese, furbearers, impounded water and recreation.
>
> The six million we got from Congress and which you think is enough, is mostly going to buy Okefenokee (Wildlife Refuge), the ranches on the winter elk range in Jackson Hole, the private lands that lie in the midst of the Hart Mt. antelope range, and for rehabilitation (dams and dikes) of the duck ranges we bought last year.
>
> By the way, Secretary Ickes wants me to give him Okefenokee. Do you mind? I don't, only that it cuts into our nesting funds.
>
> I need $4,000,000 for duck lands this year and the same bill which gave us the $6,000,000 specifically stated that at your discretion you could allocate from the $4,800,000 money for migratory waterfowl restoration.
>
> We did a good job last year. Why cut us off now?

In reply President Roosevelt wrote that he had commented to his budget director that "this fellow Darling is the only man in history who got an appropriation through Congress, past the budget and signed by the President without anybody realizing that the Treasury had been raided", and continued in his reply to Darling,

> You hold an all-time record. In addition to the six million dollars ($6,000,000) you got, the Federal Courts say that the United States Government has a perfect constitutional right to condemn millions of acres for the welfare, health and happiness of ducks, geese, sandpipers, owls and wrens, but has no constitutional right to condemn a few old tenements in the slums for the health and happiness of the little boys and girls who will be our citizens of the next generation!
>
> Nevertheless, more power to your arm! Go ahead with the six million dollars ($6,000,000) and talk with me about a month hence in regard to additional lands, *if* I have any more money left.

While at the Biological Survey, Darling was also busy in organizing public enthusiasm and participation in wildlife conservation. He was instrumental in the formation of the General Wildlife Federation and was elected its first president in 1936, then reelected president in 1937. In 1938 the General Wildlife Federation changed its name to the National Wildlife Federation. It is now the largest conservation group in the country with a membership of over 6,000,000.

Many years later Darling wrote, "Looking back, I think the approximately two years I spent with the old Biological Survey were the two most exciting years of my life." He accomplished much in those two years.

It would be incorrect to leave this section with the impression that everything is in satisfactory condition with our public lands. There is encroachment on some of our Fish and Wildlife Refuges. There are complaints as to timbering procedures at the National Forests. There are severe environmental problems threatening some of the National Parks such as the inability to see across the Grand Canyon on most summer days. New sites are added to the national register of preservation locations but no funding is included for their rehabilitation and maintenance. And the National Park Service is undermanned for all the responsibilities mandated to them.

Haiti—Deforestation

Whatever the problems in the vast lands set aside in the United States, they pale when compared to the conditions in the Caribbean nation of Haiti.

In the native Arawak Indian language, the word Haiti has been variously translated as "mountainous" or "green island". It certainly remains mountainous but it is far from green. In 1974, a septuagenarian recalled in an interview for *The New York Times*, "Port-au-Prince (the capitol of Haiti) used to be like a tropical Switzerland with pine covered hills hovering 3,000 feet above the city. Soon it will look like a slum on the moon." The old fellow was right. Haiti has lost more than 97% of its forest lands. Only a few pockets of contiguous forest parcels remain, and one-third of the nation's land has been seriously eroded.

Much of the devastation is the result of clearing of land for agricultural and grazing purposes, and this is understandable.

Haiti-Dominican Republic border. Ravaged landscape to the left in Haiti and thick forests to the right in the Dominican Republic. From Cobb, 1987. James P. Blair, National Geographic Society.

After all, much of the United States was cleared for such purposes; without it the United States would not enjoy the agricultural productivity that it now has. The problem comes when mountainous regions are cleared and much of Haiti is mountainous. Rains erode the slopes and wash away the fertile soils; the land becomes barren. In a vicious circle, farmers must cut trees on mountain slopes to grow subsistence crops but soon the rain washes away the soil and productivity falls. The farmers are obliged to move further up the slopes where the process is repeated. The dramatic effect of this continuing degradation process can be seen in the accompanying aerial photograph of the boundary between Haiti and the Dominican Republic. (Haiti occupies the western portion of the island of Hispaniola and the Dominican Republic the eastern portion.) The region to the left in the photograph is Haiti, treeless, denuded and barren; the region to the right shows the forested mountain slopes of the Dominican Republic.

Relentless and ubiquitous land degradation is Haiti's foremost environmental problem. Agricultural productivity has continued to decrease. Denuded areas are not simply degraded but destroyed. As a consequence, most people move from rural to urban areas where there are limited employment opportunities or migrate to the Dominican Republic, Cuba, and the United States. There are no easy solutions to Haiti's present problems.

The process of deforestation has been going on for generations. To understand how this has come about it is necessary to look back at Haiti's dramatic and, unfortunately, mostly tragic history.

Christopher Columbus landed on the north coast of Haiti on his first voyage of discovery in 1492. He claimed the island in the name of Queen Isabella and King Ferdinand for Spain, calling it "the most beautiful in the world…almost like the lands of Castile; rather, these are better."

On Christmas eve, 1492, with every man asleep except for an apprentice at the helm, the Santa Maria, his lead ship, went aground on a coral reef off the north coast of Haiti and had to be abandoned—not exactly the greatest feat of seamanship by the man known as the Admiral of the Ocean Sea. With the help of the local Arawak Indians the ship's stores were salvaged. When he sailed back to Spain with his remaining two ships, Columbus left 40 volunteers as the first settlement in the New World, appropriately called La Navidad, i.e., Christmas. When he returned in 1493, he found "nothing but ruins and carcasses where he had left fortifications and Spaniards".

On his second voyage Columbus established a second settlement (and the first permanent colony in the New World) farther to the east on the island of Hispaniola at the site of Santo Domingo, the present capital of the Dominican Republic. He ruled the colony as royal governor until 1499. In 1506 sugar cane was introduced and Hispaniola became the most productive colony in the New World, the envy of Europe. The Spaniards also introduced genocide to the New World. All the native Arawak Indians were enslaved and then eliminated either through small pox, a European import, or cruelty. They were replaced by black slaves from West Africa.

In 1697, the French acquired the Haitian portion of Hispaniola and renamed it Saint Dominique. By 1709 it had become the richest of the French colonies, populated by black slaves representing the dominant majority, white French colonials, and mulâtres with varying traces of Caucasian blood. Then, in 1791 began the most important series of events in the history of Haiti. The black slaves rose in rebellion against the French and, after twelve bloody years, established the second free nation in the Western Hemisphere and the world's first black republic. Their three prominent black generals were Toussaint L'Ouverture, the ideological leader; Jean-Jacques Dessalines, a fierce and vindictive soldier; and Henri Christophe, or as he preferred to be known, Henry Christophe.

By 1802 Toussaint controlled all of Haiti. But also by 1802, after an armistice with England, Napoleon, then First Counsul of France, was free to turn his attention to Haiti. He dispatched an army of 20,000 "to annihilate the government of the blacks in St. Dominque". He was partly successful; through a ruse the French captured Toussaint and sent him to France to eventually die in a prison cell. Dessalines carried on the fight and eventually the French under Rochambeau capitulated. By 1804 Haiti had become a free republic again and Dessalines declared to the people that "we must live free or die"—a motto which the people of New Hampshire will recognize from their automobile license plates. The fighting had been plagued with atrocities on both sides—a never-ending part of Haiti's history. At one point Rochambeau had 1200 unarmed black soldiers bayoneted and dumped into the harbor of the northern port city of Cap Haitien. At the end, 800 sick and wounded French soldiers left behind under safe conduct were put aboard barges, taken to sea, and drowned.

Dessalines had been a slave and his hatred for the French colonials remained unassuaged. After the departure of the French army, he ordered his soldiers to go from city to city and from plantation to plantation to systematically kill all the remaining white French men, women, and children—the second genocide in the history of Haiti. In one fell swoop all the expertise in crop cultivation, land management, exporting, and trading was gone, and the massacre isolated Haiti from the rest of the world for decades to come. All that Dessalines could offer in their place was a military dictatorship.

Dessalines declared himself Emperor, but only two years later he was ambushed and killed by a group of his own soldiers—another seemingly never ending aspect of Haiti's history. Haiti was then divided between Christophe, controlling the northern portion, and Alexandre Pétion, a mulâtre, controlling the southern portion. Christophe was a colorful figure and a builder extraordinaire. He had fought with the French on the side of the Thirteen Colonies in the American War of Independence and had been wounded during the siege of Savannah. In 1811 he declared himself Henry I, King of Haiti and established a nobility consisting of four princes, eight dukes, twenty-two counts, and thirty-four barons. He built an enormous fortification, the Citadelle Henry, which still remains and several palaces including Sans Souci, styled after Frederick the Great's palace of the same name. His nobility and strict rule were, however, short lived. In 1820, Christophe became partially paralyzed from a stroke; his soldiers revolted with the cry of "Liberty! No slavery! No king"; and he killed himself. Pétion's rule in the south was more gentile. He divided up the former French plantations and parceled them out to the ex-slaves. He died of natural causes, one of the few rulers to do so, in 1819.

Since then, Haiti's history has been a series of plots, revolutions, and military executions. Over a seventy-year period ending in the early 1900s, there were over a hundred civil wars, insurrections, and coups. Of the twenty-two presidents during this period just one served a complete term and only four died of natural causes—a devastating and unstable history.

Then, came the U.S. Marines. A largely forgotten fact of American history is that the United States occupied and controlled Haiti from 1915 to 1934 under the now departed doctrine of Dollar Diplomacy. At that time, as is true today, there was concern about European and American investment abroad, but in those days forceable takeover of a country where American interests were threatened was considered legitimate. If nothing else, the occupation did provide a period of stability with an improvement in the country's infrastructure—roads, communications, etc.

During this period a number of Haitian doctors went to the United States to study including Francois Duvalier, later to become the country's infamous dictator, known far and wide as Papa Doc. Duvalier ruled from 1957 to 1971, establishing himself as President-for-life in 1964. He passed on that title to his son, Jean-Claude, who in turn ruled from 1971 to 1986 when he was encouraged with United States support to abdicate and leave the country.

Both Duvaliers ruled by means of their dreaded secret police, the Tonton Macoute. Literally in Créole, the native language of Haiti, Tonton Macoute means Uncle Knapsack, the antithesis of Tonton Noël, or Uncle Christmas. Good children received presents from Tonton Noël; bad children were whisked away by Tonton Macoute, never to be seen again. An apt name for the Duvaliers' secret police.

The ever-present caste society and corruption in Haiti developed to an extreme under the Duvaliers and remains as striking as one will find anywhere in the world. The minority mulâtres, representing 10 to 15% of the population, own or control nearly all of the country's wealth. Their language is French; they are literate and a superbly educated elite, many having gone to school in France; and their religion is Catholic. The remaining 85 to 90%, the blacks, own little and are largely illiterate. They speak Créole, an unwritten language that is a blend of provincial French and West African dialects. Their religion is voodoo, an amalgam of animalistic cults of West Africa and Catholic ritual. The income of the country largely goes into the foreign bank accounts of the mulâtres through misappropriation and outright fraud, with little attention being given to the welfare and well being of the country as a whole. It is so unbalanced that in the poorest country in the Western Hemisphere, the dissolute thief, Jean-Claude Duvalier, spent three million dollars on his wedding in 1980.

And such is Haiti's sorry state today. There has been continuing suppression and instability. There have been few leaders who have made the effort to improve the economy of the country, to develop sustainable agricultural crops, and to introduce land management practices. Little of its extracted wealth has been plowed back into the country. Its vital statistics, dominated by the black majority, are a cold reminder of this. In 1989, population density was 580 people per square mile; literacy, 23%; unemployment estimated to be 50%; per capita income, $300 per year; male life expectancy, 51.2 years; and female life expectancy, 54.4 years. The corresponding figures for the United States are 68 per square mile, 99% literacy, 5% unemployed, over $16,000 per capita income, 71.5 life expectancy years for males, and 78.5 years for females. And the deforestation of Haiti continues as the only available opportunity for those few who remain on the land.

Brazil—Tropical Rain Forests

The ongoing and systematic elimination of the tropical rain forests of the world is one of the most talked about problems of the planet. These ecosystems contain an enormous and varied flora and fauna and they represent one of the last frontiers for development. The largest of these is the tropical rain forest of Brazil, the region drained by the Amazon River—the world's largest river, contributing one-fifth of the total freshwater river runoff to the oceans. The region, known as Amazonia, is comparable to that of the continental United States east of the

Mississippi River. An estimated 20% of all known plant and animal species live there.

The Amazon was discovered—in the sense of European discovery—by the Spanish navigator Vicente Yáñez Pinzón in 1500, not long after Columbus's voyages to the Americas. The first traverse of the river was made by Francisco de Orellana in 1542. Captain Orellana was a member of an expedition led by Gonzalo Pizarro, brother of the infamous conqueror of the Incas, across the Andes in search of a new El Dorado, city of gold. The expedition made little progress and no city of gold was found. Eventually Orellana broke off from the main party with sixty men and carried on with small boats down the river system, eventually accomplishing a 2,500 mile traverse to the mouth of the Amazon, an extraordinary feat that took a year and a half.

His journey was ably chronicled by his companion, the Dominican Friar Gaspar de Carvajal, who vividly described an encounter with some Indians along the way.

> On the following Thursday we passed by other villages of medium size, and we made no attempt to stop there. All these villages are the dwellings of fishermen from the interior of the country. In this manner we were proceeding on our way searching for a peaceful spot to celebrate and to gladden the feast of the blessed Saint John the baptist, herald of Christ, when God willed that, on rounding a bend which the river made, we should see on the shore ahead many villages, and very large ones, which shone white. Here we came suddenly upon the excellent land and dominion of the Amazons. These said villages had been forewarned and knew of our coming, in consequence they [the inhabitants] came out on the water to meet us, in no friendly mood, and, when they had come close to the Captain, he would have liked to induce them to accept peace, and so he began to speak to them and call them, but they laughed, and mocked us and came up close to us and told us to keep on going and [added] that down

below they were waiting for us, and that there they were to seize us all and take us to the Amazons. The Captain, angered at the arrogance of the Indians, gave orders to shoot at them with the crossbows and arquebuses, so that they might reflect and become aware that we had wherewith to assail them…More than an hour was taken up by this fight, for the Indians did not lose spirit…I want it to be known what the reason was why these Indians defended themselves in this manner. It must be explained that they are the subjects of, and tributaries to, the Amazons, and, our coming having been made known to them, they went to them to ask help, and there came as many as ten or twelve of them, for we ourselves saw these women, who were there fighting in front of all the Indian men as women captains, and these latter fought so courageously that the Indian men did not dare turn their backs, and anyone who did turn his back they killed with clubs right before us, and this is the reason why the Indians kept up their defense for so long. These women are very white and tall, and have hair very long and braided and wound about the head, and they are very robust and go about naked, [but] with their privy parts covered, with their bows and arrows in their hands, doing as much fighting as ten Indian men, and indeed there was one woman among these who shot an arrow a span deep into one of the brigantines, and others less deep, so that our brigantines looked like porcupines.

He further elaborated on the female warriors:

In this stopping place the Captain took [aside] the Indian who had been captured farther back, because he now understood him by means of a list of words that he had made…The Captain asked him what women those were [who] had come to help them and fight against us; the Indian said that they were certain women who resided in the interior of the country, a seven day journey from the shore, and [that] it was because this overlord Couynco was subject to them that they had come to watch over the shore...The Captain asked him how, not being married and there being no man residing among them, they became pregnant; he said that these Indian women consorted with Indian men at times, and, when that desire came to them, they assembled a great horde of warriors and went off to make war on a very great overlord whose residence is not far from that [the land] of these women, and by force they brought them to their own country and kept them with them for the time that suited their caprice, and after they found themselves pregnant they sent them back to their country without doing them any harm; and afterwards, when the time came for them to have children, if they gave birth to male children, they killed them and sent them to their fathers, and, if female children, they raised them with great solemnity and instructed them in the arts of war. He said furthermore that among all these women there was one ruling mistress who subjected and held under her hand and jurisdiction all the rest, which mistress went by the name of Coñori. He said that there was [in their possession] a very great wealth of gold and silver and that [in the case of] all the mistresses of high rank and distinction their eating utensils were nothing but gold or silver.

Not surprisingly, Orellana dubbed the river, River of the Amazons—the name by which it is still known today. There is, however, no Land of the Amazons; the story given by the Indian captive was, of course, fiction. On the other hand, eyewitness accounts—with some exaggeration—of Indian women fighting alongside men was accurate and was common in the fighting between the Indians of northern South America and the Caribbean and the Spanish.

Following the Papal accord of 1493 and succeeding treaties which divided the Americas between Spain and Portugal, the Portuguese began the gradual settlement of the coastal regions of Brazil from Belém at the mouth of the Amazon River to Recife at the eastern tip of Brazil to Rio de Janeiro and São Paulo to the southwest. Amazonia was left relatively undisturbed until the rubber boom of the nineteenth century.

Rubber trees are indigenous to the rain forests of Amazonia. They exude a white milky substance known as latex, which has the unusual, and nearly unique property among common natural materials, of being elastic. It is also waterproof and an excellent electrical resistance material. Rubber was introduced to Europe and North America in the late 1700s, for balloons, waterproof clothing and boots, and elastic bottles. Unfortunately it had a serious deficiency for much further use. It was temperature dependent, becoming brittle in cold weather and soft in warm weather.

This deficiency was corrected in 1839 with the development of vulcanization. In the presence of sulfur and heat the rubber becomes, so to speak, cured and the resulting compound remains tough and firm under both warm and cold conditions. With the oncoming industrial revolution and the demand for rubber in the automotive and electrical industries, the rubber boom was on, and Brazil was the principal supplier—a nice monopoly.

Manaus, 1,500 miles up the Amazon River and in the center of Amazonia, became a boom city. Fortunes were made and profligate spending was the norm. The most elaborate memorial to this period is the Teatro Amazonas, the Manaus opera house, which still stands today. No expense was too much—tiles from Alsace, chandeliers from Venice, marble from Italy. At its opening on 31 December 1896, an Italian opera company performed, and in subsequent years Caruso appeared on its ornate stage.

But, as with most booms dependent on a sole supplier position, there had to be an end. In the late 1800s, an English adventurer smuggled rubber seeds out of Brazil. The seeds were cultivated at Kew Gardens in London and the seedlings were sent to Singapore in 1877. They grew as cultivated rubber tree plantations. To be sure, plantations had been attempted in Brazil but the closely packed trees were destroyed by a local fungus; this was not the case for the Far East. Tapping widely dispersed wild rubber trees was no competition for the efficiency of plantation harvesting. By 1915 Asian production exceeded that for Amazonia; the Brazilian rubber boom was over. The decline was further exacerbated by the chemical development of synthetic rubber, a cheaper replacement to natural rubber for many uses.

The second boom in Amazonia began in the 1970s with the conscious decision by the then military government of Brazil to develop the region for timber, agriculture, and pasturing, and to provide a home and economic opportunity for the burgeoning population. The decision was part of an overall plan for the development of Brazil beyond its coastal cities which began in 1960 with the movement of the nation's capital from Rio de Janeiro to the interior and the building of a new capital city, Brasília.

To begin with, one can ask the question why should this not be done? After all, vast regions of Europe and North America were once covered with forests and these regions were denuded to provide land for agricultural and industrial purposes more suitable to mankind. It was a natural development and without it the Western world would not have the prosperity that it now enjoys. Why should not Brazil be allowed to follow the same course even if it means wholesale deforestation?

There are two reasons. First, there is now, which there was not when Europe and North America were developed, an environmental impetus to preserve the Earth's natural ecosystems. And the tropical rain forests are the richest and most diverse that exist. In the Western world it is now difficult to obtain developmental permits on the few remaining wilderness areas. This is, indeed, felt in Brazil and there are efforts to preserve rain forest enclaves within the developed regions.

The second reason is more economic but equally important and may, in the end, be the deterrent for continued development in Amazonia. Unlike the soils in North America and Europe, which are rich in nutrients and can support agricultural crops and grasses for grazing, the soil is poor in the tropical rain forests. Most of the nutrients in these ecosystems are bound and stored in the standing plant mass itself, not in the roots and soils. The soils

Francisco "Chico" Alves Mendes Filho. From Revkin, 1990. Miranda Smith Productions, Inc.

have few nutrients and little organic matter. Once a clearing has been made, crops may do well for the first couple of years, but thereafter the soils become impoverished and the yields are poor. For cattle grazing or crops this, then, requires the continuing slash-and-burn of forest areas, leaving behind barren and unproductive regions. The economics of such operations are not competitive with natural grazing lands that can be used year after year.

The recent development in Amazonia has much in common with the American West of the nineteenth century—life on the frontier. It is not the romanticized Hollywood version of the settling of the American West, but the lawlessness where might makes right and killing is an acceptable conclusion to a disagreement. And it also has it heroes—among them Chico Mendes. But unlike the Western movies the good guys don't win.

The antagonists are the ranchers. They are subsidized by the national government and consider that their destiny is to clear the land, graze cattle, and raise Brazil from its debtor nation status. They obtain their land for clearing either directly from the government or by negotiation and sale from the local rubber tapping community, or by outright intimidation or extortion— whatever it takes.

Amazonia has also in the same process become a home for

renegades from the more civilized and law-abiding parts of Brazil. One of the more notorious of that group is the Alves da Silva family with grandfather Sebastião, sons Darly and Alvarino, and grandson Darci. Sebastião had spent four years in jail for killing a neighbor before moving to Acre, the westernmost province in Brazil. The family along with their cowboys, sometimes serving as hired gunmen, have a history of killings often for the most trivial of reasons without ever being brought to justice. For example, one of the workmen at their ranch jokingly asked to marry Darly's nine year old daughter; he later turned up shot to death.

Chico Mendes also lived in Acre. The president of the local rubber tapper's union, he and his coworkers on several occasions confronted and stopped others from cutting the forests and successfully petitioned the national government to set aside enclaves for the rubber tappers.

Thanks to the efforts of some leading ecologists and environmentalists and the media, he and his efforts to save the tropical rain forests gained international attention. Mendes became the point man and the national spokesman for the rubber tappers and for the forests. He traveled and spoke to the United Nations and to the United States Congress as well as to other august assemblies. Back in Acre, however, he was a thorn to the ranchers and a marked man. He knew it. During 1988 four other rural union presidents in Amazonia had been murdered. On the evening of 22 December 1988 Mendes was sitting in his home playing dominoes with two body guards assigned to protect him. Mendes got up to go out the back door, where two men were waiting in ambush. With a light shining behind him from the house Mendes was killed by a shotgun blast.

Thanks to his international fame, the Mendes killing did not go unresolved. One hundred and fifty police officers were sent to investigate. The Alves clan went into hiding. As early as December 26, Darci came out of hiding and confessed to the crime. Darly was also apprehended and arrested for having ordered the killing. Darci's cohort in the ambush and Darly's brother, Alvarino, fled the area. Both Darci and Darly were convicted in February 1990 and sentenced to nineteen years in prison.

What is the future for Amazonia? Thanks to the efforts of

Chico Mendes and his colleagues, among many others, there are some hopeful signs. The Brazilian government has become more environmentally conscious. Efforts are underway to preserve at least portions of Amazonia as natural wilderness areas. Forestry management practices are becoming more prevalent rather than the chop-and-run techniques of the past. Cattle ranching may, possibly, have reached its peak for the simple economic reason that rain forest soils cannot support a year after year grass crop.

For the wild rubber tappers the future is not bright. Many have already migrated out of the forests to the villages in search of whatever jobs may exist there. Now, added to the competition from the rubber plantations of the Far East and synthetic rubber, there are rubber tree plantations in the more temperate regions of coastal Brazil where they go unattacked by the Amazonia fungus. They now supply over half the rubber from Brazil and their production will increase in the years to come.

CHAPTER 2

*…To Overextend the Limits of
Water and Land Use*

Venice—Disappearance of a Venerable City

The scholar and art critic, Bernard Berenson, called Venice "man's most beautiful artifact". It has the most renowned architecture, statuary, and religious frescoes of any city save Florence. With its canals, it may well be the most romantic city besides Paris. But Venice is disappearing—sinking while the buildings, statues, and paintings are being ruined by noxious atmospheric fumes.

Venice is, or rather was, an island; it is now connected by a causeway to the mainland. It is set in a lagoon at the head of the Adriatic Sea between barrier islands on the seaward side and Europe's largest petrochemical complex on the adjacent mainland.

Being in a deltaic region, Venice suffers from natural subsidence. That subsidence has been accelerated over the past several decades by the ground water demand for the industrial complex on the mainland. As the groundwater table has fallen, the lagoon bed has also sunk. And when onshore winds blow, the city is inundated by noxious fumes.

During the Middle Ages Venice was a major trading city and the European gateway to trading with Asia and China. Marco Polo was a Venetian. It was a City State with its leader, the Doge, chosen by the people. It is the site of St. Mark's Cathedral with the relics of the Gospel writer, Mark—albeit, those relics having been stolen by a group of enterprising Venetians from Alexandria in 829 A.D. It is the site of the Ducal Palace, home of the Doges, and the Grand Canal that traverses the city along with its tributary canals. It was home for the great artists, Bellini, Giorgione, Tintoretto, Veronese, and Titian, the greatest of all; and many of their works adorn the walls of the cathedrals and palaces of Venice. In a later era it was home for the composer and musician, Antonio Vivaldi.

From the beginning the Venetians were well aware that the proper use of water was critical to their continued existence. They established a Magistracy of Waters with supreme powers. Introducing a new commissioner one of the doges commented on his importance, as follows. "Here is a magistrate of waters. Weigh him, pay him, and if he makes a mistake, hang him."

By the late 1700s Venice had fallen on hard times. Her empire was nearly gone. She was taken by the French in 1797, ceded to Austria a year later, and became part of the newly formed Italian Kingdom in 1866. During much of this and the following times she became a carnival and gambling city—an old world Las Vegas.

Following the end of World War I, Count Guiseppe Volpi and other businessmen saw the opportunity to reinvigorate Venice by linking it to the growing industrial world. Volpi did not propose that the industrial development be within Venice itself, which would have ruined the city, but rather on the adjacent mainland. The result was the industrial complex of Porto Marghera, and its associated residential city of Mestre. The beginning industry was small and prospered under the dictatorship of Mussolini. Expansion to its present large size came after the end of World War II. Venice is the closest Western European port to the oil fields of the Middle East, the same affinity to the East that led to Venice's rise to power in the Middle Ages. In 1950, work was begun on the large petrochemical complex that is now a permanent fixture of Porto Marghera. At that time, in Venice as well as

St. Mark's Square, Venice, during a flooding event. From Judge, 1972. Photo: Albert Moldvay, National Geographic Society.

elsewhere, little consideration was given to possible environmental consequences of such an enormous facility.

But the consequences were soon forthcoming. One of the more bizarre events occurred in 1973. The fumes from the industrial facilities regularly polluted the air over Porto Marghera and Mestre and, with westerly winds, Venice as well. On 7 January 1973 Giuseppe Lo Grosso, head of the provincial labor office, decided to do something about it. He ordered the over 200 industrial companies in Porto Marghera to issue gas masks to their 50,000 workers. As unusual as the order was, it did focus attention on a problem, specifically that the fumes caused the citizen's eyes to run and throats to burn with (even as yet) uncatalogued medical damage. As one merchant in Mestre put it, "You have heard about the marble cancer (in Venice)? Well, if the muck in our air gets at marble and bronze in the city across the lagoon, imagine what it does to our lungs here." As might be expected, the industries resisted the order and encouraged the central Italian government to conduct an investigation, which in effect negated Lo Grosso's order. The problem was sidetracked and eventually nothing was done.

Meanwhile, Venice was sinking—in recent years at a rate of

one inch every five years. That may not sound like much; but when measured over the past several hundred years, it has amounted to several feet. First floor levels of residences along canals have had to be raised and then raised again. Water creeps by osmosis up through the buildings and building fronts and facades fall off or have to be propped up.

The most devastating effects of the sinking have occurred during those occasional periods of especially high tides in conjunction with sirocco, or southerly, winds blowing up along the Adriatic Sea. These high-water events have been exacerbated by the infilling of the lagoon for expansion of the industrial region and the broadening and deepening of the ship channels to Porto Marghera. The lagoonal area over which the tides can spread out has been reduced and the deeper ship channels permit a larger tidal influx to the lagoon.

Such devastating events have occurred in 1966, 1979, and 1986—the largest being in 1966. At that time St. Mark's Square was knee deep in water for a period of twelve hours.

The deterioration of Venice is obvious to even the most casual visitor. Some first floors have been closed off and boarded up, unfit for habitation. Statues have been eroded beyond easy recognition with missing noses and fingers. The interior plaster walls of churches and palaces have fallen off exposing damp building blocks beneath. And many of the once magnificent frescoes have been attacked and darkened by the industrial fumes.

So what has been done about all this—to attempt to restore Venice to its original grandeur or at least curb its continuing decay. There is both some good news and some bad news. As with many of the tales of the environment in this book, the good news and restoration efforts have come from dedicated individuals or small groups of individuals. The bad news has come from the lethargy, inaction and, in some cases, downright deterrence by the local, regional and national governments. It is an interwoven story of progress and failure.

The 1966 flood focused the world's attention on the problems of Venice. One of those most concerned was René Maheu, then director-general of UNESCO, the United Nations cultural organization. UNESCO had just finished saving Abu Simbel in Egypt, preserving it on high ground while the Aswan high dam was being built. UNESCO now addressed its attention to saving Venice.

Maheu as well as others were instrumental in seeing passage of a bill by the Italian parliament in 1973 for the city's restoration. The first sentence of the act stated that "Saving Venice and its lagoon is declared a question of the essential national interest". The act was quite specific about expenditures. Half a billion dollars would be raised and divided roughly into three parts— "one-third for the national government to repair seawalls and bridges, to combat damage done by high water and to construct barriers in the channels as a defense against high tides; one-third to the regional government, charged with devising antipollution measures and building aqueducts and a sewage treatment facility; and one-third to the city of Venice for restoration of decayed housing, palaces, bridges and monuments in the historic center and on the islands in the lagoon".

Shortly after passage of the bill a loan of five hundred million dollars, one of the largest loans arranged on the Eurodollar market, was achieved through a consortium of investment banking firms including Lehman Brothers and Kuhn Loeb in New York and N.M. Rothschild in London. The loan was entitled "A prefinancing agreement in connection with the preservation project of the city of Venice". Expectations for the restoration of Venice were never higher.

Bizarre as it may seem, *none* of the loan money ever went toward the restoration of Venice. It was used instead to prop up the lire on the world money market. George Armstrong of the *Manchester Guardian* wrote on the missing funds in 1973:

> The mystery of the missing £ 200 million which had been promised by foreign banks in 1971 to preserve the sinking city of Venice grows deeper. The official word from the Treasury Minister, Signor La Malfa, is not to worry. The Italian state will fulfill its pledge and, thus, incidentally obey a law passed last April, and the money will be forthcoming. Or, in another interpretation of La Malfa's cryptic statement, the £200 millions is already in the Bank of Italy, and

Venice will not be forsaken. The fact is that Italy *did* float a loan in London on 26 September for the same amount as authorized by the special loan. The beneficiary of that loan is now known. Signor Ferrari Aggradi, who when he was Treasury Minister, promoted a law authorizing the international loan, has said in an interview "the least we can expect is that this matter be cleared up" and that meanwhile, as to the whereabouts of the £ 200 millions, "every inference is possible". However, the only inference seemingly possible from Signor Ferrari Aggradi's words is that the money arranged for Venice has been disbursed elsewhere.

To the credit of the Italian government it had set up in 1969 an institute in Venice to study how barrages might be built at the entrances to the ship channels at the outer lagoonal islands to curb the high-water events on the city. Its first director was Roberto Frasseto, an intelligent, engaging, and world renowned oceanographer. A war hero as well, he was an Italian frogman during World War II, captured during an attack on Alexandria, and received the Medallio d'Oro, the equivalent of the Congressional Medal of Honor, after his release at the end of the war. He had studied engineering at Florence before the war and completed his education as an oceanographer at Yale and Columbia Universities afterwards.

The studies by Frassetto and his colleagues were professional and thorough. They came up with a plan for gates to be raised from the sea floor at times of high water events to prevent excessive flooding of the lagoon. The trouble was, as with many such environmental studies, that it would cost a lot of money to install such barriers—figures ranging upward of twenty million dollars. There, the project floundered.

By 1975 Frassetto had become an embarrassment to the Italian government. As is the custom in such matters, he was promoted upstairs—and out of the way. In valedictory remarks, he said:

> You cannot write off six years just like that. I will always be at the service of reasonable and honest

people and I intend to continue to make my contribution to try to save Venice. I refuse to be eliminated because I do not belong to a political current. I belong to a scientific current, and I cannot be conditioned into silence or into allowing scientific truth to be manipulated for political ends.

He went on with the following:

There is an illness in government today, and Italy has it worse than most. When a politician takes power, then he must also take responsibility. The pleasures of power are among the rewards for taking that responsibility. But in Italy the people who take power decline responsibility. This happens at all levels of government. So for those people there is only one reaction when they are faced with the responsibility of making a decision. They avoid it. They either criticize the proposal so as to justify doing nothing, or if pushed for an answer, they say no, because it involves less immediate risk not to do something than to go ahead and do it.

But there are some things that require doing straight away. Agreed, there is a risk in doing them. Every action has a risk. You can reduce this risk, by taking proper advice, but in the end you have to take a decision and pay for it if it is wrong or get the glory for it if it is right. In Italy it is hard to find anyone in government who thinks this way. It is hard to find anyone who has the power to say yes and the courage to do so. No decision can be taken without criticism, lengthy discussions, and the intervention of a great mass of ignorant people. You can't even confine the discussion to prepared people. Must equal weight be given to ignorance as to judgement and scientific skill acquired by a lifetime of preparation?

Politicians are often afraid of scientific truths. They are so locked into their own immediate interests they cannot see beyond today. They have no time to look to the future. They have no time for scientific advice about the future. They delude themselves that by repairing past damage they are saving Venice, when they need to act now to save Venice for tomorrow. They say "If we do this, what will happen in the long term?" There is only one certainty about the long term—if they continue to do nothing, Venice will die.

If the Italian government was not active in saving and restoring Venice after the shock of the flooding event of 1966, there were others who were. There have been several such enterprises and we recite here the stories of four of them. Founded by Lord Norwich and Sir Andrew Clarke, Britain's Venice in Peril successively took on the tasks of restoring the churches of San Nicolo dei Mendicoli and the Madonna dell'Orto; Sansovino's Logetta at the base of the campanile in St. Mark's Square; and the Porto della Carta, the main entrance to the Ducal Palace. John MacAndrew, a professor at Wellesley College, and Robert Maccoun, a retired engineer, were instrumental in Save Venice Inc., the United States counterpart to the British Venice in Peril group. They undertook the restoration of Santa Maria dei Gesuiti, a spectacular church which had suffered much decay. Other similar Save Venice commissions emerged in the Netherlands, Belgium, Luxemburg, Switzerland, Australia, and Iran—a truly international effort.

Venice was also the site of the first Jewish ghetto, the Ghetto Vecchio, a place which still exists and a name which is now applied universally to ethnic conclaves. (The reader will recall Shylock in Shakespeare's *The Merchant of Venice*.) The oldest synagogue in the ghetto is the Schola Grande Tedesca, begun in 1528. One of its leaders, Giorgio Voghera, undertook the task of seeing it restored and obtained the funds through a German businessman with the only stipulation being that it was to be considered as an anonymous gift and act of reparation.

The most high powered group in the restoration of Venice has been the International Fund for Monuments. It was started by James A. Gray, a retired United States Army colonel who was among the first of those landing in Sicily during World War II, later a miliary attaché in Rome. His organization started with the renovation of the stone heads on Easter Island but soon became active in Venice. He was responsible for the restoration of the great series of paintings by Tintoretto in the Scuola di San Rocco; the Scuola di San Giovanni Evangelista, a palace built in the fourteenth and fifteenth centuries; and the church of Santa Maria del Grigho. But that has not been all. The International Fund for Monuments has chapters in various cities throughout the United States. The Los Angeles chapter took on the restoration of the church of San Pietro di Castello and the Boston chapter the Scuola dei Carmini.

The work by these dedicated groups has been prodigious. By 1972, 30 palaces and churches and 1,500 paintings had been restored at a cost in excess of two million dollars. That amount may sound small in comparison with the five hundred million dollar loan that never was, but it was real money and it was well used. Even here, however, the Italian government played a dog-in-the-manger role. They demanded a 12% value added tax on all restoration funds; in other words, for every $10,000 of funds raised, $1,200 went off the top into the Italian treasury.

Finally, let us hear from the Venetians who care and who have lived through this whole troublesome period. Countess Cicogna Volpi is the daughter of one of the founders of the Marghera complex, she has written:

> My father did what he thought was right, to save Venice. I am doing the same. The survival of Venice becomes, more clearly as time goes by, the emblematic fight between two irreconcilable points of view. On the one side lie culture, tradition, history, a deep feeling that our past should be preserved, at least in its highest manifestations. On the other side we have vested interests, a blind belief in a certain form of so-called progress. No doubt, culture will be defeated unless the civilized world steps in.

Vladimiro Dorigo is an author who lives in Venice:

> What worries me is the fashionable view at the

moment—that all Venice's problems will be solved if pollution is eliminated and Venice is returned to a pre-industrial condition. That is a delusion. This is the twentieth century and people can't live in it as though it were the seventeenth century. There are structural problems—engineering problems with scientific solutions—of great complexity, and the real problem in Venice is that they are being dealt with by politicians and bureaucrats who are not up to it. In Venice these people deal with everything—industry, renovation, the arts—in an amateurish sort of way. Politicians aren't capable of handling the simplest problems of administration because of the way they are selected. They are in power because of party, money, religion, and the ability to keep quiet. The result is the selection of fools instead of technically qualified professionals.

The politicians will change, but the property speculators will go on. The tourists will continue to visit here, but the Venetians themselves will become increasingly insensitive to their surroundings. The operations designed to mend the cracks will continue, but they are not enough. Time is limited, and the men who govern Venice seem incapable of understanding that they have a deadline. If something positive happens, it will be luck, a million-to-one chance. The problems of Venice are tragic, the problems of Italy are farcical. Venice is a tragedy within a farce.

Individuals and groups of individuals have done much to preserve Venice, but they can only accomplish so much. It is now up to the Italian government to build barrages that can be raised at times of high water, institute atmospheric antipollution measures, and construct sewage treatment facilities. That may, or may not, come about.

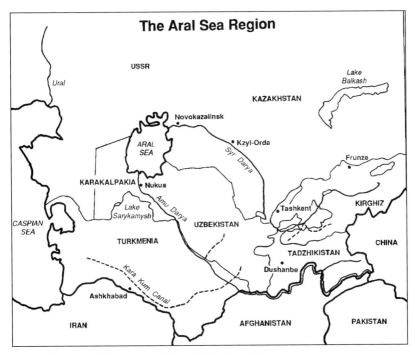

Figure 1. The Aral Sea region. From Kotlyakov, 1991.

Aral Sea—Requiem for an Inland Sea

The Venice debacle affected thousands of people; the disaster of the Aral Sea in Central, or Inner, Asia affected millions of people. The Aral Sea was once the fourth largest lake, or inland sea, in the world. Prior to 1960 it was slightly larger than either Lake Huron or Lake Michigan and somewhat smaller than Lake Superior. Then from 1960 to 1987 its surface level dropped by nearly 45 feet and its area decreased by 40%. The Aral Sea is supplied by two rivers—the Syr Darya from the east and the Amu Darya from the southeast. Its level is controlled by a balance between the river flow inputs and the desert region's high evaporation rate. Since 1960, there has been severely curtailed river inflow to the lake caused by the withdrawal upstream of water for irrigation purposes. As a consequence, the lake has been gradually drying up into a residual brine pool (see Figures 1 and 2).

Water is essential to agriculture and to mankind. When it is plentiful, there are no problems; when it becomes scarce, all kinds of economic and environmental forces come into play. For exam-

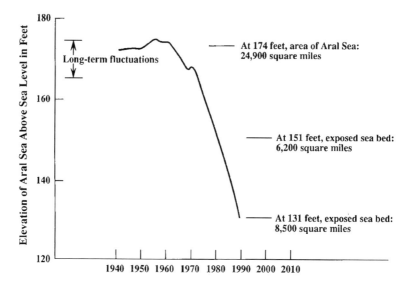

Figure 2. The shrinking Aral Sea, change in lake level as a function of calendar year. Adapted from Precoda, 1991.

ple, the once mighty Colorado River of the American West which cut the majestic Grand Canyon is now only a trickle where it enters the Gulf of California. Upstream, its waters have been drained off for irrigation and other purposes. Were the Gulf of California an inland sea rather than an arm of the Pacific Ocean, it would have dried up to a saline lake by now.

There are, however, significant differences between the case of the Colorado River and that of the Syr Darya and Amu Darya. The demands for the Colorado River waters have been going on for years with much infighting between states and between individuals and with economic reasoning being counteracted, at least in recent years, with environmental concerns. Each dam, irrigation project, and water supply diversion on the Colorado has gone through a long battle. It has not been a perfect solution, but it in no way can compare with what has happened in Central Asia.

The Syr Darya and Amu Darya diversions have been part of a massive, Soviet style endeavor which began in earnest in the 1950s. In the Soviet's centralized and state controlled government, individual concerns and the economic and environmental effect on the Aral Sea region were given little shrift. The sole purpose was to increase agricultural production in the otherwise mostly desert regions to the east and southeast of the sea. A. Babayev, a former president of the Turkmen Academy of Sciences, explained the government thinking:

> I belong to those scientists who consider that drying up of the Aral is far more advantageous than preserving it. First, in its zone, good fertile land will be obtained. According to preliminary computations this will provide for 1.5 million tons of cotton a year. Cultivation of this crop alone will pay for the existing Aral Sea with its fisheries, shipping and other industries. Secondly, many scientists are convinced, and I among them, that the disappearance of the sea will not affect the region's landscape.

It should be noted that the former Soviet Republic of Turkemenistan is within the region irrigated and the former republics of Uzbekistan and Kazakhstan are the ones that adjoin the Aral Sea. Politics as usual.

Central Asia is a remote area and little visited by tourists except for the intrepid ones who have been to Samarkand or Tashkent. Historically, it has been largely a nomad's region and a thoroughfare for thundering warlike hordes that have come variously from the east, the west, and the south. Alexander the Great extended his empire across Western Asia as far east as the Amu Darya and Syr Darya, known to him as the Onux and Jaxartes Rivers, in 329 B.C. His control of the area and that of his followers lasted for about a hundred years until the region was taken over again by the peoples of the former Persian Empire. Next came the Hun invasion from the east with their fearless leader, Attila. They started from Mongolia around 170 A.D. and by 450 A.D. had conquered all of Central Asia and much of Europe, being one of the main contributors to the demise of the Roman Empire.

Following this, particularly during the sixth century, the Turks, a branch of the Huns, settled Central Asia and their settlement became one of the largest nomad empires to that date. The Turkic people remain the dominant population of the region.

Although related, these Mongolian Turks from the east are not to be confused with the latter-day Ottoman Turks from the west and the peoples of the present nation of Turkey.

In the seventh and eighth centuries the Moslem Empire to the south expanded into Central Asia. Their religion was adopted by the Turks, peaceably and not by force; and Islam remains the religion of the region. Next came another thundering horde from the east, the Mongols with their leaders, Genghis Khan and his grandson, Kublai Khan, in the thirteenth to sixteenth centuries. The Mongols also adopted the Islamic religion.

In the following years the region was divided up into a number of loosely knit nomad empires, or khanates, until the region came under control of Russia. That control has remained until the recent demise of the Soviet Empire.

One of the decrees of the new Soviet government in 1918 was to divert the waters of the Amu Darya and Syr Darya to irrigate millions of acres of otherwise desert areas for cotton production. By the 1930s the Soviet Union had accomplished part of their goal and the nation had become a net exporter of cotton. The big change came in 1954 when construction was begun on the Kara Kum Canal. The Kara Kum, completed in 1956, diverts the waters of the Amu Darya to the western portion of Turkmenistan adjacent to Iran. The canal is 850 miles in length, essentially a large ditch with no protective lining so that a great deal of water is lost along its long course to the irrigation areas. By 1960 the combined effects of the Kara Kum and other irrigation projects began to be felt in the Aral Sea region.

As with many large and seemingly robust natural systems, you can tinker with them up to a point with no noticeable environmental effects. But once a certain threshold is crossed, things fall apart; and they do so rapidly and in ways that were not anticipated or predicted. Such has been the case for the Aral Sea and adjacent regions. The surface level has dropped by 45 feet over the past twenty-seven years and its area has decreased by 40%. The volume of the sea has decreased by 70%. By the year 2000, presuming present practices continue, it has been predicted that the Aral Sea will only be 20% of its original volume.

Along with this decrease in water volume the Aral Sea is turn-

Abandoned ships in a dried up portion of the Aral Sea. From Ellis, 1990. David C. Turnley, © Corbis.

ing into a residual brine pool with salinity having increased from 10 parts per thousand to 30 parts per thousand, a value comparable to that of ocean salinity. And extensive salt flats, former lake bottom, have been exposed on the eastern and southern shores of the former lake.

The port of Muiniak, once on the southern shore of the Aral Sea, was until recent years one of the great fishing ports of the Soviet Union. More than 10,000 fishermen there produced around 11% of the total fish catch of the nation—around 100 million pounds annually consisting of bream and barbel. Now, there is no fishing and the fishing boats are left high and dry on the former lake bed.

Along with this extensive fishing there were large canning facilities. In order to prevent further political and economic disruption, the central authorities in Moscow decreed it advisable to transport fish caught in the Baltic Sea off Latvia and Lithuania to Muiniak for processing—a distance of 2,000 miles. That makes about as much sense as transporting shrimp caught off the Gulf Coast of the United States to Toronto for processing.

The new salt flats are also a problem. They represent a sort of salty badlands. Dust storms lift the salt and spread it over the fertile deltaic region of the Amu Darya to the south. That land now produces only meager crops and more water is needed than previously to flush out the salty soils.

The climate has also changed. In the past, the Aral Sea, as is the case for all such large water bodies, provided an ameliorating effect on the climate. Now, there are greater extremes in climate, colder in the winter and hotter and dryer in the summer.

A further problem has arisen from the excess use of pesticides and fertilizers in the irrigated regions used for cotton and rice production. They have polluted the ground water and drinking water supplies and these regions have also had to be repeatedly flushed out to maintain a decent level of agricultural production.

Finally, and most tragic in human terms, the health and living standards of the people in the region have deteriorated due to the combined effects of atmospheric and ground water contamination. The incidence of typhoid has increased by thirty times over what is was previously and that of hepatitis by seven times. There is a prevalence of kidney diseases, gallstones, and chronic gastritis. And infant mortality is now more than 50 per 1,000 births.

What is to be done? One of the clearer and more perspicacious plans is that of Arkady Levintanus, a geographer at the Ministry of Nature Utilization and Environment Conservation in Moscow. First, it has to be recognized that there is a serious problem and a concept of conservation and restoration of the Aral Sea and normalization of its ecology and sanitary, medical, and socioeconomic conditions has to be accepted. The regions under cotton and rice cultivation have to be reduced to around 70% of their present extent, a draconian measure but one that is necessary to increase the water flow to the Aral Sea. Product losses due to inadequate storage and processing facilities have been high, around 20-30%; corrective measures have to be taken in this regard. The present efficiency of the irrigation systems including the Kara Kum Canal is low; modernization of the irrigation systems is required. Special payments for water uses, such as exists for parts of the Western World, should be introduced.

Will this, or anything like it, take place? The dissolution of the Soviet Union and the present unrest among the former Soviet

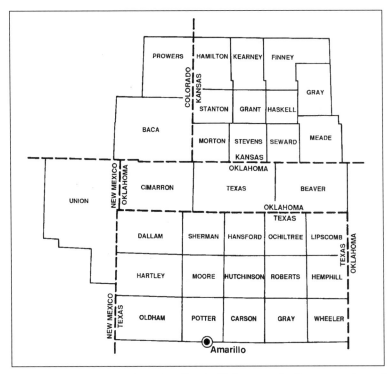

Figure 3. The Dust Bowl. Panhandle of Oklahoma and adjacent portions of Texas, New Mexico, Colorado and Kansas. From Bonnifield, 1979.

Republics does not make for a favorable situation. And the regional separation of much of the cotton and rice producing regions from that of the affected Aral Sea region with separate autonomies does not help.

Dust Bowl—Desertification

The greatest agricultural disaster that has ever occurred in the United States was the Dust Bowl of the 1930s. The migration of destitute farmers from the southern plains to California and the plights of their families have been superbly chronicled in John Steinbeck's *The Grapes of Wrath* and in the songs by Woody Guthrie. Except for the numerous displacements of Indian populations, it was the largest forced migration in American history. It resulted from a massive agricultural expansion in the 1920s into marginal land that had a history of volatile climactic changes. The delicate ecological balance of the region was destroyed with unfortunate and dire consequences. In essence the farmers pushed the land beyond reasonable limits, much as in the case of the Aral Sea.

The Dust Bowl was centered on the Oklahoma Panhandle and includes the adjacent portions of Texas, New Mexico, Colorado, and Kansas. Geographically it is referred to as the High Plains, east of the southern portion of the Rocky Mountains. As far back as the early 1800s, the great American explorer, Zebulon Pike, as well as others referred to this region as part of the Great American Desert—it is a semiarid and treeless region with, at that time, a sparse covering of buffalo grass (see Figure 3).

It is an area that is subject to climactic extremes including droughts, blizzards, and dust storms. In recent geologic times ravaging of the region by dust storms is attested to by the Sand Hills in nearby Nebraska and the broad expanses of loess soil—silt, sand and clay deposits of wind borne origin.

Dust storms result from high northerly winds blowing down from Canada across the Great Plains. The ensuing dust storms are sometimes, appropriately, referred to as "rollers" or "black northerns". The winds lift loose topsoil and form a rolling wall up to fifty feet or more in height which roars across the landscape. The larger storms are black enough to blank out the sun, and a person caught in such a roller can only distinguish objects a foot or so away. In historic times, one such extreme roller occurred on 2 May 1904. The noon train at Texhoma, a town in the Oklahoma Panhandle, was stranded for twelve hours before the storm lifted and it could continue. Another giant roller occurred on 17 March 1923, at a time when the region was already being developed by farmers. One important item that can be eliminated as a cause of the Dust Bowl is inadequate prior knowledge.

Earlier, cattlemen had tried to make a living on the High Plains. Their domain over the region extended from the middle to late 1800s. It was a marginal grazing land, at best; and in typical American fashion the cattlemen expanded grazing beyond what the region could stand, much the same as the farmers would do in a later generation. Overgrazing, collapse of the cattle market, and most particularly, the severe blizzards of 1885-1886 and 1886-1887 combined to end the cattle industry. During the first of those two devastating winters, 85% of the cattle perished on some ranches. In essence the cattlemen were bankrupted by the weather.

Then, came the farmers. It was the boom years of the 1920s, the Roaring Twenties. Everything seemed to be booming, and agriculture was not to be left out. This era on the High Plains is known as the Great Plow Up. Literally every available piece of land was plowed and planted, mainly with wheat, though much of it was land ill-suited to agriculture.

On a yearly basis little rain falls on the High Plains. The rain that does fall is either absorbed by the soil or evaporated; there is no surface runoff. The agricultural development of the High Plains during the 1920s was based on the, then, seemingly sound hypothesis of "scientific dry land farming". The object was to enhance the absorption of moisture over evaporation. The advocates of the hypothesis asserted that a dust blanket would insulate the soil and prevent evaporation; they called it a "dust mulch". To achieve this dust blanket called for new machinery. The one-way plow was introduced, a series of disks on a wide horizontal beam. The plow blades just scratched the top of the soil and left a fine powder of soil on a hard packed base. Unfortunately what was utterly overlooked was that the dust blanket also produced an enhanced source of material to be lifted into the air when the black northerns came through. In addition to the one-way plow, tractors were available for pulling the plow, combines for harvesting the wheat, and trucks for moving the produce to the market place or distribution center. All of these devices increased immensely the land that could be cultivated over the previous energy sources of animal and human labor.

In the beginning there were, indeed, boom years and bumper crops, and that only led to further development of the High Plains. In addition to those who moved and settled on farms in the region, there were "suitcase farmers" who came for the short periods of planting and harvesting but lived elsewhere, spending the bulk of their time at their separate professions as businessmen or whatever.

The harvest of 1931 was one of the largest, and also one of the

Month	1933	1934	1935	1936	1937
January	4	2	2	0	9
February	4	0	7	9	14
March	14	6	11	18	18
April	17	6	20	16	21
May	12	2	6	14	23
June	7	2	1	1	17
July	3	0	1	2	15
August	3	1	1	1	10
September	3	0	0	0	7
October	0	1	2	1	—
November	2	2	1	7	—
December	1	0	1	4	—
Total	70	22	53	73	134

Table 2. Dust storms, 1933-1937, as recorded at Panhandle A. & M. College. From Bonnifield, 1979.

last. Then things began to fall apart. There was too much wheat and not a sufficient market for what was harvested. The price dropped from a dollar a bushel to twenty-five cents a bushel. It became uneconomic for many to plant wheat. The suitcase farmers departed, leaving their land fallow with its dust cover. In the perpetual cycle of the region's weather, the rains stopped, replaced by drought. And, then, beginning in 1932, the dust storms began to roll in (see Table 2).

Drought and dust storms continued for the next five years. As shown in the accompanying table, there were not a few but many such storms each year, mainly during the spring months. One of the worst, shown in the photograph, occurred on 14 April 1935, known as Black Sunday. The dust storms lifted the dry mulch topsoil, ruined the wheat crops, and deposited the dust as a desert-like pumice elsewhere or as embankments alongside houses and barns. There was no escaping a dust storm; it came through window and door seals and covered everything with dust. Automobiles failed due to the static electricity in the air or dust carried into the engine.

The autobiography of the Dust Bowl years by Lawrence

The infamous dust storm of 14 April 1935, advancing on people near Hooker, Oklahoma. From Bonnifield, 1979. Emma Love.

Svobida gives a first hand account of living through a dust storm:

> Only those who have been caught out in a "black blizzard" can have more than a faint conception of its terrors. When the soil has become finely pulverized by too much working over, by the action of water followed by wind, or, particularly, when the surface is blow dirt from a previous storm, the dust begins to blow with only a slight breeze. As it continues to rise in the air it becomes thicker and thicker, obscuring the landscape and continuing to grow in density until vision is reduced to a thousand yards, or less. If this is to be a real dust storm, a typical black blizzard of the Dust Bowl, the wind increases its velocity until it is blowing at forty to fifty miles an hour. Soon everything is moving—the land is blowing, both farm land and pasture land alike. The fine dirt is sweeping along at express-train speed, and when the very sun is blotted out, visibility is reduced to some fifty feet; or perhaps you cannot see at all, because the

dust has blinded you, and even goggles are useless to prevent the fine particles from sifting into your eyes, though they break the force of the driving dirt.

Thus it is when the observer is within the area of a storm's inception. At other times a cloud is seen to be approaching from a distance of many miles. Already it has the banked appearance of a cumulus cloud, but it is black instead of white, and it hangs low, seeming to hug the earth. Instead of being slow to change its form, it appears to be rolling on itself from the crest downward. As it sweeps onward, the landscape is progressively blotted out. Birds fly in terror before the storm, and only those that are strong of wing may escape. The smaller birds fly until they are exhausted, then fall to the ground, to share the fate of thousands of jack rabbits which perish from suffocation.

Human beings run for their lives, if there is any safe place within reach. Some run anyway, well knowing that unless shelter is reached, they may be victims of the same fate that overtakes the birds and jack rabbits. Many harrowing tales of the Dust Bowl have been told by hitchhikers caught out in one of these storms, with no certain knowledge of the region to guide them to a place of safety…

Actually, I have seen one engine that filled up with dirt while it was standing still, though I would have found this difficult to believe if I had not had the opportunity to examine it with my own eyes. This was a brand-new tractor belonging to a machinery company in business at Meade. They had received shipment of a carload of these engines, and, as they had not enough storage space to accommodate all of them, this one was left outside in the open, where the owners knew it would be exposed to the dust-laden gales, but never suspected that these could cause any trouble with a standing tractor. It stood outside a week before it was sold, and only when an effort was made to start it was it discovered that the motor was stuck solid. The examination which followed this discovery revealed a condition that was amazing. Everything was filled with dirt. The pistons were wedged in the cylinders so firmly that, after the connecting rods had been loosened, they had to be pounded out. There was only one feasible explanation that could be offered or accepted. The dirt had been blown through the exhaust outlet with such force of wind behind it that it had been driven past the valves into the cylinder, then past the close-fitting pistons into the crankcase…

Electrical phenomena incidental to a black blizzard are many and varied. Lightning and thunder are often continuous and add to the unnerving effect of the storm, particularly in the case of people from happier sections of the country who are experiencing their first dust storm. But even residents of the Dust Bowl feel none too secure. Sometimes a filling station has to be closed down while a storm is at its height, because electricity is running back and forth through the iron pump foundations, which assume a rosy hue. There are very few attendants foolhardy enough to attempt to fill a tank with gas when these conditions prevail…

We who live in the Great Plains don't think much of the kind of country the region has become, yet we are still likely to resent the criticism of a stranger. I am reminded of a tourist from the Pacific Coast who stopped at a service station in Liberal, Kansas. While the attendant was servicing his car, the tourist talked about his trip from the West, and dwelt at some length upon the impressions he had received in driving through the dreaded Death Valley, in California. Then he talked of the part of eastern Colorado and western Kansas he had passed through, and stated with emphasis: "Why, this country is nothing but a desert!"

The filling station attendant, resentful of this

Windblown sand blocks the door of a Cimarron County, Oklahoma farm building. From Parfit, 1989.

> remark coming on the heels of the visitor's impression of Death Valley, retorted: "You went through worse desert back there in California." "Yes, that is true enough," the tourist agreed with a smile, "but there aren't any fools out there trying to farm it."

Another account but of conditions within a house during a dust storm is given by a farmer's wife as taken from the history of the Dust Bowl by Donald Worster.

> All we could do about it was just sit in our dusty chairs, gaze at each other through the fog that filled the room and watch that fog settle slowly and silently, covering everything—including ourselves—in a thick, brownish gray blanket. When we opened the door swirling whirlwinds of soil beat against us unmercifully…The doors and windows were all shut tightly, yet those tiny particles seemed to seep

> through the very walls. It got into cupboards and clothes closets; our faces were as dirty as if we had rolled in the dirt; our hair was gray and stiff and we ground dirt between our teeth.

Overexpansion into marginal agricultural regions, the collapse of the wheat market, the onset and continuation of the extreme drought and dust storm conditions, low crop yields, and the national Depression led to the mass migration out of the Dust Bowl, mainly to California. Added to this was the mechanization which favored large farm areas and fewer laborers, and the actions of the newly formed Soil Conservation Service to return parts of the region to their original grasslands. The small farmers and the tenant farmers were the ones that suffered most and made up the bulk of the migration. In Cimarron county in the Oklahoma Panhandle, 40% of the community migrated; for other Dust Bowl counties the numbers were somewhat less.

Most of the people, however, stayed and made out as best they could. As often is the case in such circumstances, humor helped ease the pain. Donald Worster repeats the often told tale "about the motorist who came upon a ten-gallon hat resting on a dust drift. Under it he found a head looking at him. 'Can I help you some way?', the motorist asked. 'Give you a ride into town maybe?'. 'Thanks, but I'll make it on my own,' was the reply, 'I'm on a horse.'" R. Douglas Hurt in his book on the Dust Bowl added the following stories:

> Indeed, people reported the drought so bad that when one man was hit on the head with a rain drop, he was so overcome that two buckets of sand had to be thrown in his face to revive him. Housewives supposedly scoured pans clean by holding them up to a keyhole for sandblasting, and sportsmen allegedly shot ground squirrels overhead as the animals tunneled upward through the dust for air. Some farmers claimed that they planted their crops by throwing seed into the air as their fields blew past and that birds flew backwards to keep the sand out of their eyes. One local report indicated that when a traveler

stopped at a Dust Bowl restaurant and ordered a boiled egg, it was full of dust when cracked. Five more eggs were opened with the same result. At this point the traveler gave up trying to eat an egg, but took a dozen eggs home and put them under a sitting hen. To his utter surprise, the eggs hatched three weeks later, producing ten mudhens and two sandhill cranes.

Economic relief finally came—for some. The 1930s was the period of development of the Hugoton and Panhandle gas fields, at that time the world's largest gas fields. They trend in a north-south direction across the Dust Bowl region. Fortunate landowners received royalties; others jobs in drilling and in the construction of pipelines to Midwest and Eastern markets. The 1930s also marked the beginning of Federal farm subsidy practices.

Since then, droughts and dust storms have occurred, but nothing comparable to the 1930s—as yet. Most of the farming now is done with deeper digging plows which bring up clods of topsoil that are less susceptible to wind erosion than the dry mulch. Some of the region has been returned to grasslands for grazing. And some of the region has been changed from dry land farming to irrigation farming. The irrigation water comes from a large aquifer, the Ogallala, which extends from South Dakota to West Texas. (An aquifer is a zone of water saturated sands and gravels at depth beneath the Earth's surface.) It takes tens to hundreds of years or more to replenish the Ogallala; it takes less time to deplete it. The Ogallala Aquifer will, by some estimates, be empty in one more human generation.

The people of the High Plains well remember the Dust Bowl era. Land and water conservation practices are prevalent. Farmers well appreciate that good years with bumper crops may well be followed by drought years with poor crops and they plan accordingly.

California and Colorado River—Water Depletion

As compared with what has happened to the Aral Sea, there is the contrasting example of another water-poor area, that of California and the American Southwest. Some take water for granted; others are daily reminded of its critical importance. It depends on where you live. We are aware of the great deserts of the world with their dearth of rainfall and paucity of life, and we do not choose to live in such regions. In the United States the East is rich in water. The West is not. And the settlement of the West both for agricultural purposes and for municipalities has been largely determined by the availability of water.

First, a little bit about the legalities of water rights in the United States before going on to the case studies of the Colorado River and the California Aqueducts. Who owns the water and who can use it? In the United States there are two systems of water rights, one applicable to the water-rich East and the other to the water-poor West.

In most of the East, water use is based on the doctrine of riparian rights. Basically, this system of water law gives anyone whose land adjoins a flowing stream the right to use the water from that stream as long as some is left for downstream landowners. It works as long as there is plenty of water available for all.

It can also lead, and has led, to procedures that one might question. The water rights of a landowner adjacent, say, to the Connecticut River are a legal entity that can be sold with or separate from the property. If someone should go along and buy up all the water rights on the Connecticut, they would, then, own the Connecticut River water and could transport it elsewhere via aqueducts to meet the water demands of New York City or other parts of the East coast megalopolis. And this is exactly what has happened in California. California, unlike the rest of the West, follows the doctrine of riparian rights, a carry over from the times of Spanish rule in the region.

In the arid and semiarid West, the riparian system does not work because large amounts of water are needed in areas far removed from major water sources. In most of the region, with the notable exception of California, the principle of prior appropriation regulates water use. In this first-come, first-served approach, the first user of water from a stream establishes a legal right for continued use of the amount originally withdrawn.

A small aside before going on to consider the specific case histories. Several years ago Robert Kerr, a cofounder of Kerr-McGee Oil Company and a long time senator from Oklahoma—an oil rich but water poor state—predicted that some day the price for a gallon of water would rise to exceed that for a gallon of oil. With some slight exaggeration one could argue that we have already reached that stage. The price of bottled water—not tap water—in food stores is around a dollar a gallon. The price of a gallon of oil at the well head is around sixty cents a gallon. (North Sea crude oil sells for around $23 per barrel and there are 42 gallons per barrel.)

The Colorado River is not the largest river in North America even though it carved out one of our more spectacular geomorphic features, the Grand Canyon. But it is the principal water source for the southwestern states. The once mighty Colorado River is now only a mere trickle when it empties into the Gulf of California. Essentially all of its waters are now apportioned out, mainly for agricultural purposes, among the states through which it flows.

The doctrine of prior appropriation worked fine until the southwest began to fill up. Then, it behooved the states, themselves, to divvy up the Colorado waters in some equitable and agreeable manner. They were also spurred to such action because of the desirability to build dams for reservoir storage of water. Those dams, particularly the large ones, would be federally funded projects; and the federal government made it amply clear that no dams would be built until the states settled their differences among themselves.

The first agreement was reached in 1922 at a series of sessions chaired by the, then, Secretary of Commerce Herbert Hoover, soon to become President Hoover. As shown on the accompanying map, it divided the waters between the so-called Upper and Lower Basins. On the basis of 25 years of data the average flow of the river was estimated to be 15 million acre-feet per year. (Now, acre-feet may sound like a strange unit to measure a volume of water for agricultural purposes. Actually it is quite sensible. Acres are the standard measure of farm land and inches of water per year the unit of measure of water needed for a given crop. Thus, acre-feet per year tells one how much land can be cultivated.) Each basin was apportioned 7.5 million acre-feet per year. The agreement specified a quantity rather than a percentage of the river flow. The agreement went on to state that the Upper Basin was obliged to deliver 7.5 million acre-feet per year to the Lower

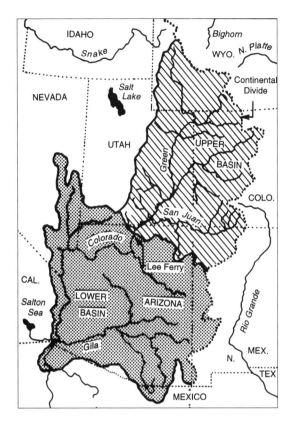

Figure 4. Upper and Lower Colorado River Basins. From Potter and Drake, 1989.

Basin—a monitored amount. Thus, in times of drought the Lower Basin got its full share and the Upper Basin got what was left. In addition, it turned out that the 15 million figure was an under estimate so that the Upper Basin suffered further.

This agreement was soon followed by the building of Boulder Dam and the formation of the Lake Mead reservoir.

The second agreement was made in 1948 among the Upper Basin states. After allocating 50,000 acre-feet per year to Arizona, the remainder was apportioned by percentages. Colorado, 51.75%; New Mexico, 11.25%; Utah, 23%; and Wyoming, 14%.

The Lower Basin states could not reach an agreement on the allocation of their 7.5 million acre-feet share. The disagreement was between Arizona, which encompassed most of the Lower Basin region, and California, which had the greater water demand. The dispute was eventually resolved by the Supreme Court in 1963 with the allocation of 4.4 million acre-feet to

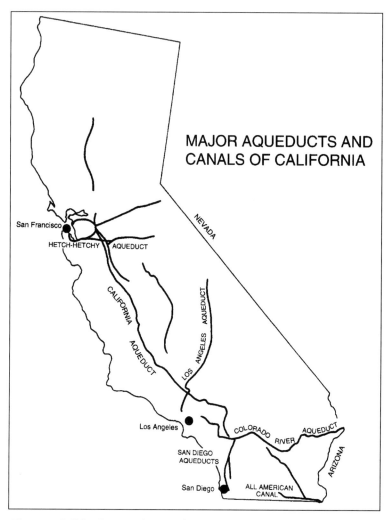

Figure 5. California Aqueducts and Canals. Adapted from Hundley, 1992.

California, 2.8 to Arizona, and 0.3 to Nevada.

This final agreement was followed by the construction of the Glen Canyon Dam and the formation of the Lake Powell reservoir. And so the wheels of progress grind slowly onward.

California has always had a water problem. It has only been exacerbated in recent years by increased agricultural production, increased population growth, and recurring droughts. The state's water resources are to the north and east and its principal users are to the south and west. California's huge agricultural industry, which grows 25% of the nation's produce, uses 85% of the state's water. The conflicting demands for dwindling water resources has pitted north against south and city residents against agricultural interests.

The key to California's growth has been available water and that water has come from the redistribution systems of the California aqueducts. A most impressive system—Colorado River Aqueduct, Los Angeles Aqueduct, and so on (see Figure 5).

Take the City of Los Angeles, as an example. The man most responsible for Los Angeles being the metropolis it is today is William Mulholland. So, who is William Mulholland? He was superintendent of the Los Angeles Department of Water and Power during the early 1900s. He saw that the Los Angeles River was inadequate to supply the needed water for a growing city. According to his estimates, the river flow would accommodate fewer than 300,000 people. He was sort of a new kind of entrepreneur—what is called a social imperialist whose goal was to acquire the water of others and to create growth in Los Angeles at their expense.

In 1904 with the approval of his board he began buying up riparian water rights in Owens Valley for the City of Los Angeles preparatory to building the first great aqueduct in California. Construction of the aqueduct began in 1908 and was completed five years later in 1913. The aqueduct had the capacity to supply the water needs of a city of 2,000,000 people—a large population number for a city back then, but not today. The Los Angeles Aqueduct has since been extended north to Mono Lake with much ill feeling in the Owens Valley and Mono Lake regions.

There was also a large piece of corruption along the way—the counterpart of insider trading on the stock market of today. As mentioned previously, the key to California's growth has always been available water. Land near a proposed aqueduct will jump in value once the aqueduct project has been announced. What happened in this case is that a group of investors including a member of the Board of Water Commissioners, the owners of the two principal Los Angeles newspapers, the president of the Union Pacific Railroad, and others bought up most of the land in the San

Fernando Valley, which was along the route of the proposed aqueduct, in 1905 prior to the public announcement of the project. And they profited enormously on the subsequent resale of the land.

CHAPTER 3

...and to Decimate or Drive to Extinction Other Species

Buffaloes and Marine Mammals— Animal Overkill

It was inevitable that mankind has been responsible for the destruction of other species. As we have grown to dominate the Earth, demanding more and more space, other species have had to give way. Some have become extinct, and many are now to be found only in zoos or, in what is nearly the same, outdoor refuges.

Virtually all the recorded extinctions of species in modern times are our responsibility. We have accomplished this in several ways. Prominent among them is outright overkill, for food, pleasure, trophies, or riches. Overvigorous hunting is the sole reason that marine mammals like Stellar's sea cow, the Atlantic gray whale, the Japanese sea lion, and the Caribbean monk seal have been exterminated. In all, overkill is the chief cause for the modern extinctions of forty-six large terrestrial mammals, as well as for 15 percent of bird extinctions, including the great auk, the passenger pigeon, and the dodo.

Predators accidentally or deliberately introduced into new places—especially that ubiquitous human companion, the rat—have had devastating effects, particularly on island bird populations. Hawaiians traveled to Polynesia bent on obtaining feathers for ceremonial purposes. The rats they inadvertently brought with them from Polynesia carried diseases that wiped out several species of Hawaiian birds. More subtle are introduced competitors, like the European starling, which was brought to New York City in the late nineteenth century and exploded outward, usurping the habitat of such birds as purple finches and eastern bluebirds, reaching by the mid-1900s into virtually every county in the United States, with an estimated population of 600 million.

Introduced diseases, from Dutch elm disease to smallpox, have devastated if not eliminated entire populations of plants and animals, including humans. Innumerable Indian tribes in the New World were decimated by the ravages of smallpox, a European disease to which native peoples had no immunity.

The human destruction of habitat has been a crucial factor in the extinction of many continental land birds and a host of other species, and this rapidly accelerating process bids fair to bring about one of the greatest and most abrupt extinction episodes in all of geologic time. How much habitat has been turned from its wild origins to human use? According to the World Resources Institute, approximately 40 percent of all the photosynthesis occurring on Earth is expended to produce energy or goods for humanity. It is estimated that Minnesota-sized tracks of rain forest are being lost each year to human activity, and there are more species of plants and animals in rain forests than in any other habitat, often highly localized species that are quickly and easily snuffed out. Often overlooked as well are other habitats such as savannas, which are disappearing at alarming rates along with species dependent on them. Species are leaving this plane of existence at an estimated rate of one a day, and the fact that most are nameless tropical insects and plants is hardly cause for sanguinity.

Migratory songbirds, for example, are showing significant declines in population as their wintering grounds in Central and South America are reduced or eliminated by deforestation and as the North American forests (their breeding grounds) are frag-

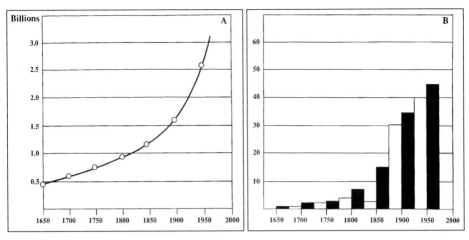

Figure 6. (A) The increase in human population and (B) the number of exterminated mammal forms (light bars) and bird forms (black bars) over the last three hundred years. From Ziswiler, 1967.

mented. If a bird population, such as that of the olive-sided flycatcher, is declining at a rate of 5 percent a year, the species could disappear well within a human lifetime.

As shown in the accompanying figure, the total number of extinctions has increased as the world population has gone up. Of the 4,200 species of mammals, an estimated 63 have become extinct since A.D. 1600, all at the hands of humans. The same is true for birds; of 8,500 species, 88 have been extinguished in the same interval. This is an extinction of one and one-half percent of mammals and one percent of birds, relatively small numbers. But if such rates (say, one percent every 400 years) continued through succeeding millennia, the extinction of one hundred percent of all mammals and birds would occur within 40,000 years. In any geologist's book, that would constitute a new and unprecedentedly rapid mass extinction. The major mass extinction events throughout Earth history have occurred over intervals of 100,000 years or so.

In geologic terms we are already in an extinction period. But unlike past extinction events, such as the demise of the dinosaurs some 65 million years ago, we do not need to search for terrestrial causes such as the climactic effects of large scale volcanic eruptions and sea level changes or extraterrestrial causes such as asteroid impacts. We know the cause—mankind.

Take the case of the American bison, or as the animal is more commonly known, the buffalo. In essence, the buffalo was doomed when the first Western settler crossed the Mississippi River. There wasn't enough land for the ranchers and farmers to graze cattle and sheep and to plant crops in coexistence with the buffalo. The well known song, "Home on the Range", with the opening line, "Oh, give me a home where the buffaloes roam", came from a poem written in 1873. By the end of the century there were no buffaloes in the Plains states and no place for them to roam.

The extermination of the buffalo was accelerated by what is known as the Great Buffalo Hunt, a period of twelve years from 1871 to 1883. During this period, the buffalo were hunted for their hides, for use in making leather goods in the Eastern states and in Europe. The rest of the buffalo, stripped of its hide, was left to rot on the prairies or to be eaten by wolves and other predators. Prior to the Great Buffalo Hunt the Plains Indians had killed buffalo for meat and clothing; but their needs were small and produced little to no diminution in the standing stock.

Prior to the 1870s the buffalo population was around 50 million head or more. That is an enormous number; they commanded the Plains. For comparison, the United States population in 1880 was the same figure, 50 million people. Many plainsmen thought that with such vast numbers of animals, there would always be herds of buffaloes. They were wrong.

It is difficult to imagine 50 million large animals occupying the Great Plains. Those who crossed the region in the early 1800s had large herds of buffaloes within view almost all the way. Numerous reports prior to the 1870s give some notion of the immensity of the herds. The following three accounts are from Wayne Gard's book, entitled *The Great Buffalo Hunt*. In 1832 Captain Benjamin Bonneville observed from a high bluff adjacent to the North Fork of the Platte River that "as far as the eye could reach, the country seemed blackened by innumerable herds". The following year J. K. Townsend while crossing the Platte Valley stopped on the rise of a hill and observed a similar scene.

> The whole region was covered with one enormous mass of buffaloes. Our vision, at the least computation, would certainly extend ten miles; and in the whole vast space, including about eight miles in width from the bluffs to the river bank, there apparently was no vista in the incalculable multitude.

In 1843 as a hunting party, chronicled by William Clark Kennerly, was about to leave camp, they saw approaching them a herd estimated to be a million animals.

> The pounding of their hoofs on the hard ground sounded like the roar of a mighty ocean, surging over the land and sweeping everything before it. Here was more game than we bargained for, and the predicament in which we found ourselves gave cause for alarm. On they came. As we were directly in their path and on the bank of the river, there was a great danger of our being swept over. This danger was averted only by our exerting every effort to turn them off in another direction. As it took the herd *two entire days* (italics added) to pass, even at a rapid gait, we were busy placing guards of shouting, gesticulating men in the day and building huge bonfires at night.

The professional hide hunters were generally young men in their late teens or early twenties. They were one of a long line of pioneers that tamed the West. They followed the missionary, explorer, and fur trader and preceded the rancher and farmer. They cleared the Plains for the settler. And, secondarily, they deprived the Plains Indians of their main source of food, clothing, and shelter, starving them into submission.

How could such a large number of animals be killed in such a few years? The killing was professional and well organized and the buffaloes were obliging, congregating in large and docile herds. Never has so great a slaughter of animals by mankind happened.

The hunting was not from horses, as shown in many panoramas of the West. It was on foot. It was accomplished with long range and accurate rifles. The rifles, themselves, were heavy and were

The buffalo hunter. From Gard, 1959. N. Eggenhofer.

often propped by a Y-stick in the ground for ease in handling and improved accuracy. The hunter would creep up on a buffalo herd, away from the wind. He would first kill the lead buffalo by a shot through the lungs. Then, any that seemed skittish and so on through the rest of the herd. The animals would simply stand around and be eliminated one after the other. Evolution and their previous experiences had not prepared them for this kind of threat.

A competent hide hunter could kill 50 to 100 buffaloes at a given stand and 2,000 to 5,000 per season. And during the peak years there were several thousand hide hunters in the field. The hunting was done systematically, first in Kansas and adjacent states, centered on Dodge City with its rail depot as the principal outfitting center and hide market. Then, in later years, south to Texas. And, finally, north to Western Canada.

There were objections to the massive slaughter, particularly from the Plains Indians, but to no avail. The prevailing opinion was that the government should encourage the hunters and ignore the complaints. The Secretary of the Interior Columbus Delano, the responsible government official, echoed this sentiment in his annual report for 1873.

> I would not seriously regret the total disappearance of the buffalo from our western prairies, in its effect upon the Indians. I would regard it rather as a means of hastening their sense of dependence upon the products of the soil and their own labors.

By the end of the nineteenth century there were only about 500 buffaloes remaining. Today, there are around 25,000 and all are on managed ranges.

The same fate has been meted out by mankind to many of the marine mammals. They are particularly vulnerable at their breeding grounds on land or on ice. We take here the case of the Caribbean monk seal. Unlike their high latitude cousins the Caribbean monk seal as well as the Hawaiian and Mediterranean monk seals spent their entire lives in tropical waters. They gave birth to pups on the beaches of the Caribbean and Bahamas. As with the buffaloes, they had no experience with predators on land. They had no fear of humans and this made them easy prey for club-wielding hunters.

Their first encounter with Europeans was on the second voyage of Columbus to the New World in 1494. On a small island south of Haiti, some of his crew came across eight seals resting peacefully on the beach. The crew approached cautiously. The animals showed no alarm or attempt to escape and all eight were killed. It was a prophetic beginning.

From the seventeenth to the nineteenth centuries they were hunted extensively for the oil that could be produced from their blubber. Two early accounts describe these hunts and the ease with which they were carried out—one in 1675 on Alacranes reef, just north of the Yucatan Peninsula in Mexico, and the other in 1707 in the Bahamas.

> Here are many seals…the Spaniards do often come hither to make Oyl of their Fat; upon which account it has been visited by English-men from Jamaica, particularly by Capt. Long: who having command of a small Bark, came hither purposely to make Seal-Oyl.
>
> The Bahama Islands are fill'd with Seals, sometimes Fishers will catch one hundred in a night. They try or melt them, and bring off their Oil for Lamps to these Islands.

In later years the few remaining Caribbean monk seals were killed by local fishermen who considered them competitors for fish. In addition, the fishermen and their families gradually took over the beaches, depriving any remaining seals of their breeding grounds. The last verified sighting of a Caribbean monk seal was in 1952. They are now considered extinct.

Passenger Pigeon and Dodo—Bird Overkill

There are numerous examples of extinctions of birds due to mankind and his activities. Many are related to small bird populations isolated on islands but the largest extinction of perhaps any species was that of the passenger pigeon.

Three hundred years ago, the passenger pigeon was the world's most abundant bird. Their population has been estimated to be three to six billion and they accounted for one quarter of all land birds in North America. They wintered in Louisiana and the other southern states and their spring and summer breeding grounds were in the northern states extending from Wisconsin across to New England.

They travelled north in flocks of hundreds of millions, filling the sky for days and blocking out the sun—shades of Alfred Hitchcock's movie, *The Birds*. They nested in colonies consisting of millions of birds, making them easy prey for mankind as had also been the case for the buffaloes. In their nesting grounds every branch of every tree could be covered with up to a dozen nests.

They were constantly on the move, necessarily because of their large numbers and the limited food supply in any given area. Their food was acorns, beechnuts and chestnuts during the fall, winter and spring and fruit and berries during the summer. Even in their nesting grounds they stayed little more than a month; then the adults and young birds, by then as large as the adults, were off again. One such description of a roosting is as follows.

> The pigeons checked their flight and settled down on the largest limbs of the tallest trees, beginning about 5 o'clock in the evening and continuing until dark to fill every tree until every available inch of space on the limbs was occupied, those arriving toward the last often flying against those already in possession and knocking them from their perch. … As far as the eye could see the air was filled with the flying birds, not in flocks but a steady downpour of feathered life. There did not seem to be any diminution in the velocity of their flight, or lessening to the height in the air at which they were traveling until they were within a few rods of the earlier arrivals and then a downward swoop with distended wings. The nearness of bird to bird, and their continuous arrival resembled the pouring of a sheet of water over the incline of an apron in a dam across a stream.

Because of the large number in any given colony, what few predators might be present in a given landing region made a negligible diminution in the colony population, that is until humans came along with enhanced harvesting techniques.

To settlers, the pigeons were an enormous nuisance, and for that reason alone their extinction was probably foreordained. The weight of the birds bent alders flat to the ground and they left forests destroyed by their sheer numbers and crops ruined:

> It sometimes happens that having consumed the whole produce of the beech trees in an extensive district they discover another at the distance perhaps of sixty or eighty miles, to which they regularly repair every morning, and return as regularly in the course of the day, or in the evening, to their place of general rendezvous, or as it is usually called the roosting place. These roosting places are always in the woods,

and sometimes occupy a large extent of forest. When they have frequented one of these places for some time, the appearance it exhibits is surprising. The grass is covered to the depth of several inches with their dung; all the tender grass and underwood destroyed; the surface strewed with large limbs of trees broken down by the weight of the birds clustering one above another; and the trees themselves, for thousands of acres, killed as completely as if girdled with an axe. The marks of this desolation remain for many years on the spot; and numerous places could be pointed out where for several years after, scarce a single vegetable made its appearance.

Although passenger pigeons had been shot while in flight or in their nesting areas for years, the wholesale slaughter of the birds occurred in the 1860s and 1870s. They were a reasonably large bird, much the same size as a squab, and were harvested for food for the Eastern markets. The modern technologies of the day—the Eastern railroad system and the telegraph—facilitated the slaughter. The birds could be easily shipped to market and the telegraph alerted the thousands of pigeon harvesters to the locations of the roosting colonies. (The birds seldom occupied the same roosting site from one year to the next.)

Adults were captured mainly by netting. The ground near a roosting colony was baited with food and underlain by a net. Vertical stakes with horizontal beams were stuck into the ground. At the far end of the beam was a small stool to which a captured pigeon was tied by its feet. A vertical rope was attached to the beam and carried over a limb to a blind where the hunter waited. The beam and pigeon were pulled up and then let to fall, giving the appearance of a pigeon landing for food. When the roosting colony took the ruse and flew down to the baited area, the overhanging part of the net was tripped and the birds captured. (Hence, the expression "stool pigeon" which we now use as a description for similar human undertakings.) A good netting could capture 100 to 200 birds. The netting operation is described in the following narrative:

Taking passenger pigeons caught in a stool pigeon, trap net and caging them for transport to market. From Blockstein and Tordoff, 1985. Bettmann Archive.

Presently, a scattered flock of some two or three hundred appears. We both sally out, and when we think (it) near enough, toss our flyers into the air. They go up the length of their lines, fifty or sixty feet, and find they are anchored, and return to the ground, wherever their blinded lot may light them. Then we rush in and "play the stool"—pulling on the cord and lifting it from the ground where it rests on a small pod of grass.

We lift it about three feet and let it drop instantly. In this operation, the stool flutters on its way downward, imitating pigeons feeding on the ground, when other flocks are passing. Soon we begin to see the flock beginning to sail, they whirl, sail over the bed, turn and sail for alighting. We never wait a second. As soon as we think we have a fair amount of them alighting and about to alight, we surge on the spring pole and spring the net, rush out and hold down the sides, to keep them in, for with their united effort, they carry the net off the ground, and the ones near the sides escape.

The fat nestlings were more prized than the adults and were easier to catch. One simply went through and chopped down the

The dodo. From Greenway, 1958.

trees to harvest the fallen birds, used poles to knock the young out of the nests, or sometimes set fire to the trees to cause them to jump out. This way, it was possible to harvest nearly all the young in a colony.

Unable to replace themselves, the pigeons were doomed just as is the case with forest clear-cutting; unless the saplings and some adult trees are left, there will be no future forest. And as their numbers decreased, they were certainly proportionally more subject to natural predators. The large colonies disappeared in the 1880s and the last passenger pigeon died in a Cincinnati zoo in 1914.

The disappearance of the dodo is probably the most widely known example of an avian extinction by mankind. It was also the first historical record of a bird species extinguished by human activities. The dodo had two characteristics which, like other similar species, made it vulnerable. It was restricted to a relatively small region, specifically the island of Mauritius in the southwestern Indian Ocean, and it had evolved into an unusual form which made it an easy prey.

The dodo was an extraordinarily gross and ungainly creature, weighing as much as 50 pounds. We do not know what the dodo's ancestors were like or where they came from many millennia ago, but what they found on Mauritius was an island devoid of predators and with plenty of food on the ground. There were no mammals other than bats. Flight became redundant and was lost, the wings becoming meaningless stubs. Its size increased with the abundant food supply and it developed terrestrial habits.

It was, so to speak, king of the roost. It had no fear and became an easy prey to humans. Hence, the expression, "dumb as a dodo".

The European discovery of Mauritius was in 1507 by the Portuguese. In succeeding years ships stopped to replenish their supplies with meat from the dodos, but the extinction of the dodos began in earnest with the settlement by the Dutch in 1598 and later development of a penal colony there in 1644. The settlers harvested dodos. They also introduced pigs, cats, and rats to the island which wrought equally great damage to the eggs and the young.

The dodo became extinct around 1680. Its demise lives in history in the familiar expression, "dead as a dodo". Its cousin, the solitaire, which lived on the adjacent islands of Réunion and Rodriguez in the Mascarene Islands, became extinct around 1800. With the settlement and clearing of the islands and the introduction of alien species, other bird populations have become extinct. As of 1967, only 21 of the original 45 bird species on Mauritius had survived, and yet others have become extinct since that date.

Gypsy Moth and Zebra Mussel—Introduced Species

Species have gone extinct, and indeed whole ecosystems have been altered by the introduction of alien species. For the United States, the number of introduced species is innumerable, measured in the thousands. Some of these introductions have been through natural pathways but most have been the direct result of human activity, both the intentional and the inadvertent.

The devastation caused by some of these nonindigenous species has been enormous—and continues to today. They vary from plant pathogens, to insects, to plants, to mollusks and fish, and finally to birds and animals. We give here a short but representative catalog.

The chestnut blight, a fungus destructive of the American chestnut, was introduced into North America at the port of New York on diseased, ornamental nursery stock from China in the

1890s and 1900s. At that time the chestnut was the most economically important hardwood in the Eastern forests of the United States, the dominant tree in vast areas of forests composing up to 25% of the trees in some forests.

It took some time for the chestnut blight to spread but by 1930, virtually all the hundreds of millions of chestnut trees had been killed. It may have been the largest single plant decimation in the history of mankind. No longer "under the spreading chestnut tree" could "the village blacksmith" stand (though Longfellow's tree was actually a horse chestnut).

A close second to the chestnut blight as a plant pathogen is the Dutch elm disease, caused by a fungus carried by the bark beetle. The disease begins with a wilting of the younger leaves in the upper part of the tree and then is carried to the lower branches. In midsummer the leaves turn yellow and fall off. Within a period of a month or so the tree is dead.

The disease was first observed in The Netherlands in the early 1900s, hence its name. It spread to the United States along with its carrier, the European bark beetle, in the 1930s on diseased timber imported from Europe. Initially its impact was restricted to the region near New York City but soon extended over the entire Northeast and is now found as far north as Wisconsin and as far west as California. It has devastated the Northeastern elms, essentially eliminating the vast numbers of shade trees, an aesthetic loss to many towns and cities as well as an expense to replace the forty million or so elms that have been lost. Some control has been accomplished by cutting and burning the diseased trees. No longer can one speak of "Desire under the Elms", and "Desire under the Oaks" just doesn't have the same ring to it.

Another introduced pest is the boll weevil. It is a beetle which feeds on the silky fibers inside the cotton bolls, or seed pods, of the cotton flower. A native of Central America, it crossed the Mexican border into Texas in the 1890s and by the 1920s had infested all the cotton growing areas of the Southeastern states. The economic losses are the highest of any harmful nonindigenous species in excess of fifty billion dollars from 1909 to 1949. Fortunately, recent eradication programs have been successful. By the 1980s, the boll weevil was no longer resident in North

Carolina and Virginia. An arsenal of weapons has been necessary. Insecticides are used on the dormant population during the winter, bait traps and chemical pesticides during the spring and early summer before they have a chance to reproduce, and sterile male releases during the breeding period.

The Mediterranean fruit fly, or medfly, is another alien troublemaker. It looks much the same as the common house fly, but infests many fruits and vegetables. The female fruit fly drills holes into ripe fruit on the tree and lays her eggs. When the eggs hatch, the larvae eat their way through the fruit which eventually drops to the ground. Considered a native of the Azores Islands, by the 1900s it had appeared in Brazil and Hawaii. In 1929 it was discovered in Florida and a quarantine was ordered. By 1930 this first infestation had been controlled. Again in 1956 the medfly invaded Florida and was eliminated by 1960.

More recently there have been economically ruinous infestations of medflies into California, each of which has eventually been controlled at considerable expense and loss of crops, as well as public concern over wholesale spraying of neighborhoods, a concern that became a hot political issue. The first introduction of the medfly into California is thought to have come from tropical produce mailed from Hawaii; more recent infestations are thought to have arrived with fruit imported from Guatemala and Argentina.

The nonindigenous European gypsy moth presently causes the greatest damage to American trees of any parasite. The gypsy moth caterpillar can chew the leaves of a wide range of trees although its preferred food is the oak leaf. In severe cases it can leave groves of trees completely defoliated. For the year 1981, the Forest Service of the U.S. Department of Agriculture estimated losses in excess of seven hundred million dollars and for the past several years the Forest Service expenditures for suppressing the gypsy moth have been in excess of ten million dollars annually.

Interestingly the introduction of the gypsy moth to North America can be traced to a single individual and a single incident. In 1869, Etienne Trouvelot brought gypsy moth eggs from France to his new home in Medford, Massachusetts. He was an amateur entomologist and hoped to develop a silkworm for America. The effort failed but he accidentally released some of the moths. Within a few years the trees in Medford began to be defoliated by the gypsy moth caterpillar. They encountered no natural predators and spread rapidly. By 1930 the entire Northeast was infested. Since then, it has spread south to North Carolina and west to Michigan. Presumably hitch-hiking on the long distance transportation of household goods, firewood and the like, the gypsy moth has spread to the entire West Coast.

Present eradication programs are much the same as they have been in the past—burning the egg masses, banding the trees with burlap or sticky material to trap the larvae or prevent their ascending the trees, spraying the trees with an insecticide, and the introduction of parasites that prey on it. At best, the gypsy moth can be contained today; it could have been eliminated completely had appropriate and known corrective actions been taken in the 1890s and 1900s. A former Chief of the Bureau of Entomology, L. O. Howard, commented as follows in 1930:

> It is a pity that the State appropriations were interrupted in 1901. It is a pity that the Federal Government did not take hold at the start and make every effort to exterminate the pest while it was still confined to the vicinity of Medford. But the government did not do things of that sort at the time. Appropriations were small and were hard to get. The economical New Englanders were tired of the expensive fight, and it is hard to blame them. Knowing what we do now, it would seem that the Federal Bureau of Entomology might fairly be blamed for lack of foresight in not warning Congress and other States of the great danger and in not appealing to Congress for funds with which to prosecute radical work. As I look back, the idea seems never to have occurred to us. It seemed to us a State matter which Massachusetts could handle if she would. There is no doubt that prior to 1901 large areas had been so carefully gone over by State forces that the gypsy moth was exterminated locally, and

we argued that if this would be done over a number of square miles, it could be done over 400 square miles then occupied by the insect.

Today the United States spends about seven million dollars a year in the attempt to manage an aquatic weed with a name that could be from a Japanese monster movie—hydrilla. Native to South America, it was introduced into Florida for use in aquariums. It is thought to have been released into canals near Tampa in the 1950s and some time later into the canals near Miami. It now infests a large number of Florida ponds and lakes and has spread to the adjacent southern states.

Hydrilla forms a dense mass across the surface of a given water body impeding lake navigation, blocking irrigation and drainage canals, and generally restricting public water uses. It is currently managed in part with herbicides, with the introduction of sterile grass carp that feed on the hydrilla, and by mechanical removal to provide waterways for boat travel. (It should be noted, however, that hydrilla blooms are just one form of cultural eutrophication—resulting from an oversupply of anthropogenically derived nutrients to a restricted water body. Were the hydrilla completely removed, other noxious blooms might take their place.)

The sea lamprey is a parasitic creature native to the North Atlantic and North Pacific Oceans. Strictly speaking, it is not a fish but more like a large worm; they may grow to as much as three feet in length. Its natural life cycle is to spawn in fresh water streams, where they die, with the young proceeding back to the oceans. Sea lampreys attach themselves to fish, scrape a hole through the skin, and suck out blood and other body fluids, eventually leading to the death of the fish—not a particularly enchanting sort of life history to contemplate.

The sea lamprey first appeared in Lake Ontario in the 1880s. By the 1920s it had migrated beyond Niagara Falls via the Welland Canal to Lake Erie, reaching all the Great Lakes by the 1930s. These later generation lampreys no longer migrated to the oceans but remained as adults in the Great Lakes and preyed on the local fish. By the 1950s, the lampreys had devastated the native fish population of the Great Lakes including lake trout, whitefish and chub.

Fortunately, through the joint efforts of the United States and Canada, a successful effort took place to reduce the lamprey population significantly and, in turn, to restore the native fish population. Two particular methods have been of importance. First, electric weirs, or fences, are placed across the streams to prevent the adults from moving to their spawning ground. Second, chemical compounds known as lampricides, which kill the larval lamprey without harming other fish, are used in the streams to destroy the young. About ten million dollars is spent annually on these control measures; without them it has been estimated that the losses to the commercial and sports fishing industries would be around five hundred million dollars.

The Asian clam is another nuisance species of alien origin. It was first noticed in British Columbia in the 1920s. When it was introduced to North America is uncertain. It is used as a food in China and may have accompanied immigrants to the United States and Canada in their water supplies. For many years the Asian clam was restricted to the West Coast, but by the 1960s it had spread eastward and is now found in most of the eastern and southern states. Its eastward migration is considered to be the result of mankind's unwitting activities rather than natural dispersal mechanisms.

The clam reproduces rapidly. It adheres to the pipes of electrical power plant cooling systems and to irrigation and municipal water systems, producing massive numbers of shells that clog the flow of water. During the 1960s and 1970s the Asian clam caused the shutdown of numerous electrical power generating facilities for corrective maintenance with down time costs of several million dollars. Since then and for unknown reasons, its populations have begun to decline.

The zebra mussel is an import from Europe. It entered the Great Lakes in the 1980s through discharged ballast water from commercial shipping, and has spread rapidly to the Hudson, Susquehanna, Mississippi, and Illinois River basins. Similar to the Asian clam, the zebra mussel reproduces rapidly and adheres to water pipes, clogging power plant and municipal water facilities.

Finally, we come to the starling, a belligerent, noisy and generally unattractive bird. An avian Attila the Hun, it evicts rightful residents, plunders fields and orchards and, in great flocks, swoops into towns and cities to wreak havoc. One can admire the intelligence and aggressiveness of starlings but it is difficult to like them.

They were introduced into the United States from their native Europe in 1890 by Eugene Schieffelin who brought over 80 birds and released them in Central Park in New York City. Not satisfied with his first effort he released another 40 in 1891. Such a procedure would not be permitted today but no such regulations existed in 1890. What possessed Schieffelin is difficult to understand; starlings were known at that time to be a nuisance species in Europe. It has been suggested that he had a crazed scheme to introduce into this country all the birds mentioned in the plays of Shakespeare.

Starlings have now spread throughout the United States at an estimated population of 600 million birds—roughly three starlings for every household cat. They are a strong and muscular bird and their aggressive behavior has driven songbirds such as wrens, bluebirds and finches from their native habitat. Starlings are, unfortunately, here to stay.

Through folly, misadventure, miscalculation, the weight of sheer numbers and in other ways, humanity has driven species over the edge of extinction and continues to do so at an ever increasing rate. Only heroic effort can halt what will otherwise surely be a mass extinction unrivaled (at least in its rapidity) in the history of the planet. And every species now, or later, found on the list of endangered species is yet another canary in the global mine.

CHAPTER 4

*But There are Hopeful Signs
and Corrective Actions have been Taken
Largely through the Efforts of
Individuals—on a Local Scale*

Love Canal—Lois Gibbs

Hardly a day goes by these days without one or more reports on the latest or some continuing environmental problem. Some are local, some regional, and a few are global. Oil spills in Alaska, photochemical haze over our cities, waste disposal in nearly every community, abandoned coal mine fires, eutrophication, acid rain, global warming, depletion of natural resources, endangered species, famine, population explosion, and the list goes on. It can, at times, be too much—too much to comprehend and too much to assimilate. Sometimes we look the other way. Perhaps the problems will disappear; perhaps they are overstated in the media and, indeed, that is sometimes the case. But, in the end with careful and rational consideration, we know that they will not go away and that we must confront and deal with them in order to preserve our own heritage and that of humanity in general for future generations.

Fortunately, many environmental problems have been addressed or are being addressed with success. In most instances, the impetus toward a resolution of a given environmental problem has come from individuals—men and women—and not from government or industry though, to be sure when sufficiently prodded, government and/or industry has become a necessary partner. Some of these individuals are environmental scientists or engineers but most are simply citizens who have been affected by a particular disaster or who have become concerned about a particular environmental problem.

To begin, we consider the story of Love Canal, an area within the City of Niagara Falls in western New York. Love Canal had been a dump for toxic chemical wastes; it was then covered with clay and became the site for a school and home construction. It was one of the first and certainly the most publicized cases of toxic waste disposal.

In the 1890s a man named William Love had the grandiose idea of building a waterway around Niagara Falls to harness the power of the Niagara River to produce hydroelectric power. His plan was short lived, leaving behind a canal 3000 feet long by 60 feet wide by 10 feet deep within what was then a farming community. In the subsequent years Love Canal provided a recreational area for swimming and fishing.

In 1942, Hooker Chemical and Plastics Corp. obtained permission to dispose of wastes in the Canal, eventually purchasing the Canal and adjacent land five years later in 1947. The company continued to use the site through 1952, disposing of more than 21,000 tons of chemical wastes, many of which were known to be dangerous. During those years it was a relatively easy matter to obtain a dump permit and little to no attention was given as to what types of material were dumped. In many ways Love Canal was an ideal site, located in a thinly populated area and lined with walls of impermeable clay.

But there were forewarnings of things to come as shown in the following three recollections from the period between 1943 to 1952. (These quotations and those following have been taken from the book on Love Canal by Adeline Levine or from the report on interviews of the Love Canal residents by Martha Fowlkes and Patricia Miller.)

They dumped all year early in the morning...my husband was home sick and the smell was so bad that we could hardly breathe and we had to put wet towels over my husband's mouth and nose. It came toward the houses like a white cloud and killed the grass and trees and burnt the paint off the back of the houses and made the other houses all black. I know for sure they did it all in '43 because my husband...died October 28, 1943 with lung trouble and I think the gases that came this way helped shorten his life...

Workers would run screaming into her yard when some of the toxic chemicals they were dumping would spill on their skin or clothes. She remembers her mother washing them down with a garden hose until first aid could arrive...

Well, I remember them dumping the chemicals. I saw them. Gee, that was shortly after I got out of the service. We used to swim in that canal, we used to fish in there, before they were dumping. It was pure water then. It was muddy, but it was pure water. I never thought about it when they started dumping. Yeah, I saw the barrels going in. They'd just back up with the trucks. It was full of water at the time, so when they dumped, it would go right into the water. ...Yes, neighbors complained. We belonged to the fire company over here. We'd go over there and put fires out. It was a regular garbage dump, too, papers and stuff like that, 'cause I can remember all that stuff burned, old wood, besides the chemical plant's stuff. It used to catch on fire. They probably put sodium in there, which always catches fire easily. But there was so much fields and open space, you didn't notice it, it was isolated...

By 1953 Love Canal was nearly full. Hooker covered it over with clay and grass and weeds grew on its rough surface. That same year Hooker sold the Canal to the School Board of the City of Niagara Falls for a token payment of one dollar. Among other items the conveyance deed had the following disclaimer.

> Prior to the delivery of conveyance, the grantee herein has been advised by the grantor that the premises above described have been filled, in whole or in part, to the present grade level thereof with waste products resulting from the manufacturing of chemicals by the grantor or its plant in the City of Niagara Falls, New York, and the grantee assumes all risk and liability incident to the use thereof. ...as a part of the consideration...thereof, no claim, suit, action or demand...shall ever be made...against [Hooker] ...for injury to a person or persons, including death resulting therefrom, or loss of or damage to property caused by, or in connection with or by reason of the presence of said industrial wastes.

One might judge that this was a good deal for both parties. Hooker got off the hook, so to speak, for any possible future problems from the toxic wastes, and the School Board got a large tract of land for free. But how any sane minded School Board could then build a school, which they did, over the filled in Canal is difficult to understand. To be sure, we are writing with twenty-twenty hindsight; even so, the School Board's actions are still difficult to understand.

Construction of the school began in 1954 and was completed in 1955. The new school accommodated 400 children. The remainder of Love Canal not set aside for the school was sold by the School Board to developers who built single family residences. As one might expect, there were construction difficulties with the building of the school. The architect advised the School Board after construction had started that the contractor had excavated a "pit filled with chemicals" and that there were two adjacent sites filled with chemical wastes in fifty-five gallon drums. He advised that it would be "a poor policy" to build on this site because of the odors and possible damage to the concrete foundations from the chemicals. The School Board ordered that the site be moved somewhat to the north of the original site and construction was continued to completion.

As construction of the residential community began and continued around the new school, new home owners arrived who had little to no knowledge of what lay beneath them. They assumed that it was safe because *they* had built a school there and *they* had given the home owners low cost FHA mortgages—*they* being the authorities, the guardians of the public trust.

Numerous incidents followed. Streets were put through; sanitary and storm sewers and utility lines were installed, breaching the impermeable clay walls of the former dump site. Construction work had to be stopped from time to time when chemical drums were exposed. Sump pumps in basements came up with an oily black substance. Huge puddles appeared in some backyards. Children who played barefoot in the fields sometimes came home with irritated feet. Rocks found in the fields would sometimes explode if they were dropped or thrown—a veritable mine field.

Yet little happened during the intervening years until 1976, when as an ancillary result of a separate investigation by the Department of Environmental Conservation of the State of New York, the Love Canal waste disposal site was rediscovered. News articles began to appear in the *Niagara Gazette*, describing the past history of the waste disposal site and its sale to the School Board and suggesting possible danger to the people living in the area.

For the next seven years from 1976 to 1983, Love Canal was front page news. A cause célèbre among environmental mishaps, it brought out the importance of citizen action, in this case the Love Canal Homeowners Association. And it brought out both the goodwill and, at the same time, the extraordinarily inept bungling and callousness of the state and federal bureaucracies. Love Canal was a morality play acted out over an extended period of time—act 1, 1943-1952, chemical waste disposal; act 2, 1953-1956, school and home construction and neighborhood living over the former dump site; and act 3, 1976-1983, concern for human health and ultimate relocation of many residents.

From beginning to end concerns at Love Canal centered around such questions as how toxic is toxic; what quantities of what chemicals were injurious; how long an exposure and through what means, ground leakage, air breathing, or what, were harmful; and who was potentially at risk, only those with houses directly over the buried dump or those at some distance away but still in the general neighborhood of the Canal? Underlying all these human health considerations was money; who was to reimburse the residents of the condemned houses; and just how many houses were to be considered within the hazardous zone? These were not easy problems to resolve at the beginning and, by and large, they still remain unresolved. As one former resident put it,

> But I think that's the whole summary of the Canal. Everybody knew what was going on and when you got right down to it, nobody knew what was going on.

The geography of the situation is shown in the accompanying map of the Love Canal area. Love Canal, itself, is the stippled portion of the diagram, encompassing about 16 acres. The neighborhood was divided into three rings and referred to as such by all parties involved. Ring I, not shown in the diagram, included Love Canal, the 99th street school, the houses on the eastern side of 97th street, and the houses on the western side of 99th street. Ring II included, as well, the houses on the western side of 97th street, the eastern side of 99th street, and the western side of 100th street as well as the adjacent houses on Colvin boulevard. Ring III extended the area out to a broader reach to the west, south, and east of Love Canal (see Figure 7).

During the two years from 1976 to 1978, things began to happen, thanks largely to the media coverage. The New York State Department of Health conducted a public health study and the City of Niagara Falls hired a consulting engineering firm to survey and design a program for reducing the pollution at Love Canal. And the residents joined in an effective protest and civic action group which became known as the Love Canal Homeowners Association.

Lois Gibbs was an unlikely protest leader. She was a young mother and a homemaker; her husband was a chemical production worker. They lived two blocks from the Canal. She became concerned when her young son began attending kindergarten at

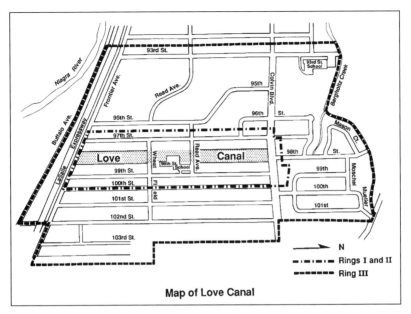

Figure 7. Map of Love Canal region. Original Love Canal, shaded portion, along with the outlines of Rings II and III. From Fowlkes and Miller, 1982.

the 99th street school. He developed asthmatic symptoms and was subject to convulsions; she wondered whether these ailments might be related to the toxic wastes at the school site. With accompanying letters from her doctors she requested in June, 1978, that her son be transferred. Her request was denied on the grounds that there was no hazard at the school.

That galvanized her into action, eventually leading through a sequence of sometimes painful learning experiences to her national prominence as a defender of people's rights in environmental crises. With her co-leader, Debbie Cerrillo, she went door to door talking with people about her concerns and encouraging them, not always with success, to join in a protest group. The Homeowners Association had three objectives—(1) restitution for property value losses, (2) clean up of the waste site, and (3) closing of the school.

The engineering survey, completed in August, 1977, found that 21 of 188 homes near the Canal had residues in their basement sump pumps and associated strong odors and that storm sewers to the west contained the harmful compound, polychlorinated biphenyl, PCB. Their study of the fields adjacent to the school

> ...revealed that areas of the Love Canal containing drummed residues have corroded drums at depths of 3 feet or less with many being exposed at the surface. Organic material from the drums is sometimes visible on the surface or is encountered within a few feet of the surface...malodorous fumes from the leaked materials are perceptible at all times.

In due course in the following years a construction project was carried out to contain the wastes and restrict the polluted area to the Canal itself. A drainage system was installed around the Canal to divert the leaking wastes to a central holding region at the southern end.

The commissioner of the Department of Health, Dr. Robert Whalen, gave his official pronouncement on the health situation on 2 August 1978. He stated that

> The Love Canal Chemical Waste Landfill constitutes a public nuisance and an extremely serious threat and danger to the health, safety and welfare of those using it, living near it, or exposed to the conditions emanating from it, consisting, among other things, of chemical wastes lying exposed on the surface in numerous places and pervasive, pernicious and obnoxious chemical vapors and fumes affecting both the ambient air and the homes of certain residents living near such sites.

Whalen recommended:

> (1) that pregnant women living at 97th and 99th Streets and Colvin Boulevard temporarily move from their homes as soon as possible; (2) that the approximately twenty families living on 97th and 99th Streets and Colvin Boulevard arrange to relocate temporarily any children under two years of age as soon as possible.

He also recommended that residents near the Canal avoid using

Compound	Acute Effects	Chronic Effects
Benzene	Narcosis Skin irritant	Acute leukemia Aplastic anemia Pancytopenia Chronic lymphatic leukemia Lymphomas (probable)
Toluene	Narcosis (more powerful than benzene)	Anemia (possible) Leukopenia (possible)
Benzoic acid	Skin irritant	
Lindane	Convulsions High white-cell counts	
Trichloroethylene	Central nervous depression Skin irritant Liver damage	Paralysis of fingers Respiratory and cardiac arrest Visual defects Deafness
Dibromoethane	Skin irritant	
Benzaldehydes	Allergen	
Methylene chloride	Anesthesia (increased carboxy hemoglobin)	Respiratory distress Death
Carbon tetrachloride	Narcosis Hepatitis Renal damage	Liver tumors (possible)
Chloroform	Central nervous narcosis Skin irritant Respiratory irritant Gastrointestinal symptoms	

Table 3. Some of the more important chemicals identified at Love Canal and the human biologic hazards associated with them. From Levine, 1982.

their basements and eating anything from their gardens. At a later date another official of the Department of Health added the following information.

> Concerning your notion about examining pregnancy data by five year time intervals, it is important to note that during the 35-40 year period during which pregnancies were counted, there were only 80 live-births and 25 miscarriages, 11 birth defects and 13 infants with low birth weights. [Among residents of wet-area homes, east of the canal, south of Colvin

Boulevard.] As you can appreciate, distributing these rather small numbers over seven or eight time intervals makes difficult interpretation.

Albeit a small sample, these are most alarming numbers. In his report Dr. Whalen also included a listing of hazardous chemicals identified at Love Canal and their possible toxic effect, as repeated here in the accompanying table. On the same day as Whalen's pronouncement the City of Niagara Falls announced that it would close the 99th street school.

As well meaning as Whalen's recommendations were, they only caused further concern—indeed, panic and alarm—among the residents. Nearly all the residents used their basements for one reason or another—laundry, recreation room, extra bedroom, etc.; what was one to do, block off part of your own home as a hazardous area? How could one be sure that the contamination and health danger was only confined to 97th and 99th streets and adjacent portion of Colvin boulevard; what about the residents of Ring III? And, then, who was going to pay for the recommended moves?

All of a sudden, after a long period of smoldering, things erupted. On the evening of the following day, 3 August, some 600 Love Canal residents met with state and local officials. It was a bedlam. One of the Health Department officials asked Gibbs to quiet the crowd so that the meeting could continue. She went to the microphone and said,

> Hi, everyone knows me, I'm Lois Gibbs, and they've asked me to calm everyone down. I suggest that you sit quietly and listen to the questions and answers and then boo the hell out of them.

There were cries of "where's Carey [then Governor of New York]", "where's the mayor", and "where's Hooker".

Indeed, on 7 August, Governor Hugh Carey did visit Love Canal. He agreed to move the residents from Rings I and II at State expense—a total of 239 families. When someone asked about those outside the selected area, he replied that they would not be included. On the same day the Federal government came into the picture. President Carter declared that an emergency existed at Love Canal and the Senate approved the requisition of Federal aid for that purpose.

In 1979, the U.S. Justice Department brought a suit against Hooker Chemical on behalf of the Environmental Protection Agency (EPA); in 1980, the State of New York joined in a similar suit against Hooker. As part of their brief, the Justice Department considered that they needed additional public health data. EPA contracted with Dr. Dante Picciano, then at Biogenics Corp., to conduct a blood chromosome study of Love Canal residents. Blood chromosome damage is an important test as to whether an individual has been exposed to toxic chemicals. It is also an indicator of possible cancer and birth defects. The study was to be a limited one of only 36 individuals with no control group for comparison. The report was to have far reaching effects—far beyond the original intention.

The completed study was delivered to EPA on 15 May, 1980. The summary statement in the report contained the following.

> We believe that the results of this pilot study indicate that these residents may have increased frequencies of cells with chromosome breaks and marker chromosomes, especially ring chromosomes. …In the absence of a contemporary control population, we have estimated that the frequency of individuals with supernumerary accentric fragments to be approximately one (1) in one hundred (100) in normal individuals. This frequency is compared to eight (8) individuals with supernumerary accentric fragments in a total of thirty-six (36) Love Canal residents studied.
>
> It appears that the chemical exposure at Love Canal may be responsible for much of the apparent increase in the observed cytogenetic aberrations and the residents are at an increased risk of neoplastic disease, of having spontaneous abortions and of having children with birth defects. However, in absence of a contemporary control population, prudence must be exerted in the interpretation of such results.

If correct, the results were indeed alarming.

Two days later, the report was leaked to the media and the panic button had been pushed one more time.

Love Canal became front page news again. This time EPA representatives came to the site to calm the citizens. Two of them arrived at the Homeowners Association office on Monday, 19 May. There was an angry crowd of 500 people outside that wanted the EPA people detained until some action came from Washington. Mind you, this was not a group of rabble rousers; these were normally peaceful citizens who went about their life in a lawful manner. But they were fed up with the bureaucracy, both State and Federal, and wanted to be relocated from what they considered was a hazardous and unhealthy region. Gibbs did detain the two in her office and called the White House to state that "the Love Canal residents are holding two EPA officials hostage". No reply came from the White House but she did receive a call from Congressman John LaFalce, who had been working on behalf of the Love Canal residents, that he was having dinner with the President that evening and would tell him that the EPA representatives were being held hostage and that the residents were tired of being told that their health was at risk with no bureaucratic action as to a resolution. Shortly thereafter she received a warning and countdown from the FBI to release the hostages. For the safety of all she decided to oblige and went out and told the crowd,

> Here is the message we should deliver to Washington. Here are your EPA people. What you have seen us do here today will be a Sesame Street picnic in comparison with what we will do if we do not get evacuated. We want an answer from Washington by noon Wednesday!

On Wednesday, 21 May, the EPA announced that temporary relocation of approximately 700 families from the greater Love Canal area had been approved. In October of the same year temporary was amended to permanent relocation by President Carter. By the summer of 1981 more than 600 families had moved.

In the aftermath to all this, the chromosome report was found to be faulty and of dubious value.

But before going on to the several reviews that were carried out, it is necessary to set the stage. When a given entity, be it government agency, academic institution or industry, wants a report slanted in a particular direction, it can pretty well assure itself of this by whom they pick to do the study. The Justice Department wanted a health report that would be favorable to their suit against Hooker.

Picciano was a controversial figure. As an employee of Dow Chemical, he had carried out a study of possible chromosome damage to workers exposed to benzene. He concluded that the workers had excessive chromosome aberrations. Dow sent the chromosome slides to an outside expert who could not substantiate Picciano's results. When Dow refused to release the report, Picciano resigned. He then went to work for the Occupational Health and Safety Administration (OSHA). They used the Dow report in a hearing on allowable exposure to benzene. The report and slides were again sent to an outside expert who, again, found the study lacking. The court decided to reject the OSHA request for lower allowable benzene exposure levels.

Over the month following the submission of the report there were five reviews. Two were at the instigation of EPA and they were negative. Three were at the instigation of Picciano and they were generally positive. One of the EPA reports received considerable media attention, particularly in the nationally circulated magazine, *Science*, from which the following remarks are taken.

> But when the EPA panel looked at the data, they saw nothing that could by any stretch of the imagination be called supernumerary accentric chromosomes. ... The EPA panel concluded that there was no evidence that the Love Canal residents had excessive chromosome abnormalities and that supernumerary eccentric chromosomes exist only in the mind of Picciano.

Those are rather strong words from one scientist to another. Finally, a sixth review at the request of Governor Carey was carried out by a blue ribbon panel chaired by the eminent doctor, scientist and writer, Lewis Thomas. That report was also negative.

On the basis of the Thomas report, the Federal government subsequently decided that it was safe to live in the Ring III region and set up a program to revitalize the neighborhood by offering the condemned houses for sale with tax abatements and substantial mortgage subsidies. The project has met with only modest success. As reported in *The New York Times* of 15 July, 1982, the announcements were made by Dr. Clark Heath of the U.S. Department of Health and Human Services and Richard Dewling of the EPA. Heath made the following statement.

> "Outside of the canal site itself and the two rings of houses around it", he said, "the Love Canal area is as safe for human habitation now as are other parts of Niagara Falls and, indeed, as safe as other residential areas in industrial towns around the country."

He was followed by Dewling, apparently unnerved by the hostile crowd—The Homeowners Association had become expert at that game by now—with the additional comment.

> Mr. Dewling said the decision to declare the neighborhood habitable came late because of extensive "negotiations". He then withdrew the word negotiations amid laughter at a crowded news conference in a downtown hotel ballroom.
> "We don't negotiate science, we discuss," he said afterward.

Later, Dewling put his foot in his mouth again.

> After their news conference, the Federal officials moved to City Hall to face local residents. "We are saying the area is habitable," Mr. Dewling said.
> "Is that safe?" cried people scattered around the room.
> "The word safe is an individual choice word," Mr. Dewling said. "There's no such thing as totally safe or zero risk."
> At another point, Margaret Bates, a former resident of the area, questioned the report's comparison

of Love Canal with other neighborhoods in Niagara Falls.

"You're saying all of Niagara Falls is contaminated," she said.

"That's a fact of life in the world we live," Dr. Heath said.

Several years later in 1988 the U.S. District Court found Hooker Chemical liable for the clean up of the Love Canal waste site and ultimate disposal of the wastes, in spite of the disclaimer in its transfer of the property to the School Board.

Times Beach—Judy Piatt

The chain of events at Times Beach parallel those at Love Canal. It is another example of the environmental and public health effects associated with the disposal of toxic wastes—in this case, dioxin. Times Beach became front page news in 1983, a few years after the events at Love Canal. There are many similarities but also significant differences between the two.

Times Beach is, or rather was, a small town in Missouri about 30 miles west of St. Louis. It is located on the banks of the Meramec River. The *St. Louis Star Times* bought the parcel of land in 1925 and sold plots for a modest price of $67.50 to anyone who agreed to subscribe to the newspaper—hence the name, Times Beach. An unusual method to increase newspaper circulation.

Here, we enter the alphabetic jungle of present day, chemical engineering—DDT, dichlorodiphenyltrichlorethane; PCB, polychlorobiphenyl; CFC, chlorofluorocarbon; PCDF, polychlorodibenzofuran; and TCDD, tetrachlorodibenzodioxin, or more simply, just dioxin. The very names of these long chain molecules synthesized by industry are enough to cause alarm. And, indeed, Times Beach emphasized the scientific uncertainty as to how harmful some of them can be.

As was the case for Love Canal, the Times Beach episode is associated with the efforts of a single individual—Judy Piatt. In 1971 she owned a large horse stable and arena. She contracted with Russell Bliss, a salvage oil dealer, to spray the stable area to keep the dust down. Almost immediately after the spraying, her animals and her two daughters became ill. Four days later, she gathered up hundreds of dead sparrows. Eleven cats and four dogs that frequented the stable died. Eventually 65 of her horses died or had to be destroyed.

She suspected that something in the oil that had been sprayed on the dirt floors was the cause. She followed Bliss around and identified 20 or so other sites that had been sprayed, including Times Beach. She ascertained that Bliss had included not only waste motor oil but also sludge from the North Eastern Pharmaceutical and Chemical Company in his spraying. The sludge, containing dioxin, was a waste by-product from the manufacture of herbicides.

She sent this information to state and federal authorities in 1973, expecting that some action would be taken, but not much happened. In 1974, the Center for Disease Control (CDC) identified dioxin as the toxic agent at the Piatt stable and at three other stables. In 1975, CDC recommended that two home sites to which contaminated dirt from one of the stables had been taken be excavated. The State of Missouri rejected the recommendation in part because dioxin, at that time, was thought to have a half-life of one year. It was later determined that the half-life was much longer, 10 years. Again, little happened until December, 1982. At that time CDC recommended the evacuation of Times Beach on the basis of elevated dioxin levels.

In late February, 1983, Anne Burford, EPA Administrator, announced that EPA had agreed to buy out the 2,400 residents of Times Beach at an estimated cost of $33 million.

Was such a hasty and precipitous action called for? Let us look at the facts as they were known at the time. The dioxin level at the Piatt stable was 30,000 parts per billion (ppb). Since the time Bliss had sprayed, the streets of Times Beach had been paved. In 1982, the dioxin level on the road shoulders there was 100 ppb and 1-5 ppb for the soils at the homes, 300 to 30,000 times less than that at the Piatt stable. CDC had established a safe dioxin level of 1 ppb. Bliss had distributed 18,500 gallons of sludge, or around 130 tons, with an original dioxin level of 300,000 ppb over a wide area—in a concentrated form at the Piatt stable and in a more dilute form at Times Beach. In contrast, at Love Canal a school and residential community had been built over a contained disposal site of 21,000 *tons* of hazardous waste material.

What happened at the Piatt stable was a tragedy. Fortunately, her daughters' symptoms disappeared after the dirt was removed. No illnesses at Times Beach could be attributed to exposure to dioxin and subsequent medical studies confirmed this. The only cases of known dioxin poisoning have been related to industrial accidents and to a chemical plant explosion in Italy which spread a dioxin cloud over the neighboring town. In these instances the illnesses consisted of anorexia, diarrhea and chloracne, a severe and disfiguring form of acne. There are no documented fatalities from dioxin. The CDC's safe-level figure of 1 ppb was not based on empirical data; it was a calculated figure.

In retrospect, it appears that the Times Beach evacuation was not necessary. It was a political decision prompted, at least in part, by the travails within EPA at that time which was under the gun for the handling of the Superfund, designed to clean up the country's major toxic waste sites. Nevertheless, the bureaucracy, once set in motion, bumbled on. In 1990, a 10 year project estimated to cost $200 million was set in motion to clean up Times Beach and 27 other sites by incinerating the soil.

Cuyahoga River—Ben Stefanski

Love Canal became the symbol for all that could go wrong with the disposal of toxic chemical wastes. Times Beach is a reminder that overreaction to such threats are possible. But, the Cuyahoga River at Cleveland, Ohio, had earlier become a symbol for a more pervasive problem—the disposal of municipal and industrial wastes.

On 22 June, 1969, the river caught on fire—actually caught on fire—and burned for six days, threatening in the process two wooden railroad bridges. One observer, Mike Tevesz, professor of geology at Cleveland State University, summarized succinctly the impact of the incident.

> The river burned in the early 1950s, too, but television news was a novelty then. But when it burned in 1969, it burned on newscasts all over the world.

The Cuyahoga fire, bizarre as it was, became a major impetus in the campaign to put environmental problems in the forefront

Boat caught in flaming Cuyahoga. Cleveland Plain Dealer.

of our national concerns. In the following year, the U.S. Environmental Protection Agency was created.

The Cuyahoga River flows southwest to Akron and then nearly reverses direction and flows northwest to Cleveland where it exits into Lake Erie. Cleveland and Akron are in the heartland of industrialized America. By the 1880s and 1890s, Cleveland was a center for oil refining with J.D. Rockefeller, steel mills with Mark Hanna, and shipbuilding with Samuel Mather. Akron was a center for rubber with B. F. Goodrich and the American cereal industry with Ferdinand Schumacher. All the wastes from these industries were simply dumped into the Cuyahoga and as early as 1900 the Cuyahoga was already polluted, a muddy brown color with low dissolved oxygen levels. Fish species such as muskellunge, walleye and lake trout had been replaced by lower quality forms, such as carp and perch.

By 1969, the Cuyahoga had taken a further turn for the worse. In the vicinity of Cleveland the river was now an oily black color with floating sludge, timbers and other debris—just waiting to be ignited. It was devoid of any visible forms of life, even such lowly creatures as leeches and sludge worms. The steel mills that lined the Cuyahoga—Republic, Jones and Laughlin, and U.S. Steel—processed 75% of the Cuyahoga River flow through their plants, primarily for contact cooling. Urban and suburban Cleveland had grown and the existing sewage treatment plants had not kept up with the growth. Many of the plants were outdated and plagued with breakdowns. In addition, Cleveland had some 600 storm sewer runoffs. In metropolitan areas with all their buildings and paved streets it is necessary to have drains to collect and transport rainwater; it cannot soak into the ground as in rural areas. The drains normally carry the water to the municipal sewerage system. However, in the case of heavy rains the storm sewer runoffs come into play and carry the rainwater *and* sewage directly to the receiving water body—in this case, the Cuyahoga or Lake Erie—and bypass the treatment plants.

As Cleveland was in 1969 a symbol of indiscriminate municipal and industrial waste disposal, it has since become a symbol of constructive and corrective actions that can be taken to alleviate such a situation. This has come about partly through the will of the citizens of Cleveland and partly through the enforcement regulations of the EPA.

Much of the credit for the improvement of the municipal waste disposal system should go to the efforts of Ben Stefanski, a lawyer who had in 1969 been recently appointed Director of Public Utilities for Cleveland. With the outlandish picture of the Cuyahoga fire in the public mind and the support of the newspapers, citizens groups, and the suburban communities, he cajoled the city into raising a $100 million bond issue directed toward building a large and modern treatment plant, upgrading existing facilities, and installing trunk sewer lines. The bond issue was to be paid off by increasing the sewer tax rate. On the industrial side Republic Steel and Jones and Laughlin have also spent about $100 million for water pollution controls. U.S. Steel closed its plant. Most of their remaining waste waters now go to the municipal system rather than directly to the river.

In 1978, nine years after the fire, phosphorus effluents, a major contributor to the eutrophication of Lake Erie, had been cut in half. Chemical pollutant discharges had decreased by a third. And oil spills had decreased by a factor of ten. The Cuyahoga River in the vicinity of Cleveland is not as pristine as it was in colonial days and it never will be. But it has returned to conditions similar to those in 1900 with muddy brown water and fish, albeit only the likes of carp, perch and goldfish. Pleasure boats, condominiums, and restaurants now line the river banks—a far cry from the fire hazard of several years ago.

As a further environmental and conservation measure, the river valley between Akron and Cleveland was set aside in 1974 as the Cuyahoga Valley National Recreation Area. It is run by the National Park Service and, as of 1978, had authorized appropriations of around another $100 million for land acquisition and park development. Not too bad a recovery for a distressed area.

Centralia—Joan Girolami

Like Times Beach, the town of Centralia in eastern Pennsylvania is largely now a ghost town with deserted streets and demolished houses. Most of the former residents took advantage of a Federal buy-out offer in 1984 and chose to move to a new community. The reason for the abandonment of the town was an underground coal mine fire which led to ground subsidence in some parts of the town and to noxious fumes, particularly carbon monoxide and carbon dioxide, seeping into the basements and thence to the living quarters of many of the houses and threatening the lives of the residents.

Fires in abandoned coal mines are more common than one might presume. There are an estimated 250 such fires, burning uncontrolled, in the United States. Most of them are in remote areas that cause little concern but not all of them fall into that category. In the 1960s an underground fire at Laurel Run, Pennsylvania, led to the removal of all of the town's 850 residents and demolition of their houses and of the other buildings in the town. But the abandonment of Centralia in 1984 was the largest and most expensive of such removals.

Centralia is located along the core of the Appalachian Mountains in the heart of the anthracite, or hard coal, mining region. Its subsurface is honeycombed by a maze of abandoned coal seams at various levels, some of which extend up to within a few feet of the ground surface.

The fire started in 1962 and was declared uncontrollable in 1983 after some sixteen separate efforts to put it out had failed. It is a story of bureaucratic bungling and buck passing—at the local, at the state, and eventually at the federal level.

The Centralia mine fire can be likened to a forest fire in the American West or a bush fire in Australia. If the fire is noticed early, it can be easily controlled. The longer it goes on, the more difficult it is to control until eventually, in some cases, it has to be left to burn itself out. In the case of Centralia, from the very beginning, the efforts were always too little and too late.

These futile and insufficient efforts at controlling the fire are well summarized by two local coal miners and a former mayor of the nearly abandoned town.

> They could have put it out in a week if they had known what to do, if they really cared about doing the job right. But nobody ever did…
>
> The coal company that used to own that land went out of business. It was an abandoned stripping when it all started. My God, it was terrible. There was supposed to be a trench to cut off the mine fire. They started digging it out here, past my backyard, went up one hundred feet and quit. They quit for lack of funds, fifty thousand dollars short. That was 1969.
>
> They had it out when they started digging the first time. All they needed was to dig another shift. They were only digging one shift a day. They should be digging three shifts a day when they're digging a mine fire. They always did before that, and since that, on every other mine fire job. What the hell. They had it out and it came to Labor Day. They had it dug right out, at the corner of the cemetery up here, and they laid off for five days. I went down in that pit and looked at it, and I could see the fire swirling this way…
>
> People always kept coming up with one idea or another to put out the fire, but it was always a day late and a dollar short. Nobody ever treated this like

a real priority. We've been lied to so much, and this is our government we're talking about...

Mining of the Centralia coal seams began in the 1840s and the town, itself, was incorporated in 1866 largely through the efforts of Alexander Rae, a mining engineer and superintendent of mines for the Locust Mountain Coal and Iron Company, which later became part of the Lehigh Valley Coal Company. On 17 October 1868, he was ambushed and murdered by the Molly Maguires while traveling between Centralia and Mount Carmel.

The Molly Maguires were an Irish-American terrorist organization that committed acts of destruction and murder in the name of protecting the miners' rights. The name, itself, derives from a legendary Irish heroine who led the Irish farmers in revolt against the absentee English landowners.

The reign of the Molly Maguires was short lived. Their leaders were brought to trial and convicted in 1875 and 1876 largely through the efforts of a Pinkerton agent who infiltrated the organization. It was at that time and perhaps still remains the most famous case of the Pinkerton Detective Agency.

A fair share of the blame, however, must be placed on the coal companies. The working conditions in the mines were abysmal and the wages, most of which went to the company store, were paltry. The miners worked long hours, lived in squalor, died hungry, and their children lived without hope.

Between the Mollies and the coal companies, it was war. The state and federal governments had abrogated their authorities. A private corporation, the Philadelphia and Reading Railroad, hired a private detective agency, the Pinkertons; a private police force, the Coal and Iron Police, arrested the Mollies; and private attorneys from the coal companies prosecuted them. The state provided only the courtroom, the jail, and the hangman.

One of the other victims of the Molly Maguires from Centralia was the Reverend Daniel McDermott, pastor of St. Ignatius Roman Catholic church. In 1869 he had denounced the Mollies from the pulpit. They were quick to retaliate, assaulting him while he was praying in the cemetery. The priest managed to get back to the church where he summoned the parishioners by ringing the church bell. When they arrived, he is alleged to have told them that from that day hence there would be a curse on Centralia and that the day would come when only St. Ignatius would remain standing in the town. A little over a hundred years later his curse became a reality; St. Ignatius is one of the few remaining buildings in the town.

Coal mining continued in the Centralia area until the stock market crash of 1929. At that time Lehigh Valley Coal was forced to close all five of its Centralia mines, putting thousands of miners out of work. During the ensuing depression some of the miners turned to bootleg coal mining. Entrance shafts were dug down to the coal seams from house foundations in Centralia. A common form of such mining was known as "pillar robbing". Pillars of coal are left at regularly spaced intervals under normal mining practice to support the ceiling over the mined area. When the pillars are subsequently removed, as in the case of bootleg mining, the mine roofs often collapse leaving the rooms filled with rubble and debris. These collapsed areas only added to the great difficulty in access when trying to control the subsequent mine fire.

Some miners returned to regular work in 1935. By that time, strip mining of the shallow coal seams had come into vogue. In the case of Centralia, strip mining left a large pit close to the Odd Fellows Cemetery. This pit, subsequently used for trash disposal, would be the site for the onset of the Centralia mine fire. The pit was 300 feet long by 75 feet wide by 50 feet deep. Strip mining often sliced through old mine tunnels and shafts and this was no exception for the Centralia pit. Several holes in the walls and floor of the pit appeared.

In May 1962, the Centralia town council decided to clean up the accumulated trash in the pit by setting it on fire. The fire department was there to put out the fire after the trash burning. They presumably did so but the fire kept smoldering and flaring up, and in the following weeks as they tried to put the fire out by bulldozing the remains, the firemen made the discovery that there was a wide hole that led to the underlying mines. Thus did the Centralia mine fire start. (From then on for the next twenty-two years there was a sequence of missed opportunities to put it out. Cost estimates went up year by year as the fire spread underground—from less than $1,000, to $10,000, to $100,000, to $1,000,000.)

In July 1962, a local contractor with experience in mine fires contacted the Pennsylvania Department of Mines and Mineral Industries and told them that he could dig out the nascent fire for about $175. He was told that his request would have to go through the bureaucracy at a time when quick action was imperative; nothing happened. In August, a local strip mine operator offered to take on the mine fire at a minimal cost as long as he could keep any coal that was recovered. He was refused. In October, the state approved a project to dig out the mine fire at a cost of $27,000. By the completion of the project, the fire was still out of control, having progressed down the steeply sloping coal seam.

Later in 1962, another project was initiated. Crushed rock and water were to be pumped down into the mine ahead of the fire to smother it and prevent it from further advancement. The project ran out of the appropriated funds of $42,420 by March 1963 and was terminated prior to completing its objective.

And, on it went. In the following years the United States Bureau of Mines and Office of Surface Mining became involved. In April 1969, with the fire continuing on its own path, a contract for $518,840 was awarded to install a fly ash barrier ahead of the fire. Fly ash is a waste product from coal-burning electrical power plants. It could be blown into the mine and presumably fill up the crevices and cavities left from the pillar-robbing subsidences.

The fly ash barrier was in due course installed and that hopefully was the end of the spread of the fire. Unfortunately the fire burned around the barrier through the low grade coal at the top of the mined out areas. Estimates for further control work were now in the millions of dollars.

By 1981, as the fire spread beneath the town, its effect could be felt above ground. In February, Todd Domboski, a twelve year old boy, went out into his backyard and saw smoke rising from the ground. Curious, he went over to examine it. The ground gave way and he collapsed into the hole. Fortunately, he was able to grab a tree root and his cousin who was nearby was able to pull him out of the steaming hole. As it turned out, the mine maps of Centralia showed that there was an abandoned mine shaft at this location which had been filled with debris. Hot steam from the mine fire had moistened and softened the debris until under its own weight and that of Domboski it collapsed.

Subsidence, smoke and fumes from the Centralia mine fire along Route 61. From Jacobs, 1986. Renée E. Jacobs

In March, John Coddington was overcome by carbon monoxide and carbon dioxide in his bed. He was barely able to alert his household and was rushed to the hospital where he recovered. Two years previously he had been obliged to close his gasoline station, on the same property as his house, when a temperature of 132° Fahrenheit was measured near the tanks. It was subsequently discovered that a bootleg mine shaft, dug by the previous owner, ran down from his basement to the coal seams.

Route 61, which was the main road through Centralia, had to be closed from time to time because of ground subsidence and when effluent smoke and fumes from the fire restricted visibility. At one point less than six feet of cover lay beneath the highway and the mine chamber. Boreholes there measured a temperature of 770° Fahrenheit—a veritable inferno lay beneath Centralia. Gas detectors had already been installed in a number of homes. In March 1981, the Office of Surface Mining agreed to pay the expenses for the relocation of thirty families in the most critical part of the mine fire impact zone.

Well before 1981, the citizens of Centralia had become distrustful of the inadequate efforts to put out the mine fire and of the misstatements and general lack of information from the state and federal bureaucracy. At one point they had been told that there was nothing to fear from the fire because there was 70 feet of rock between the fire and the ground surface which would prevent any leakage of mine gases. That was patently false. At another time a Federal Bureau of Mines official had the following brief exchange with a resident at a public meeting.

> "If you had put out the fire when it started, you wouldn't have this," the resident rejoined.
> "I don't want to argue with you," Kuebler said. "As I previously told this group, this is your fire and not the Bureau's."

The general question of who was responsible for the fire and who was responsible for putting it out remained a thorn throughout the whole episode, impeding proper control efforts.

In 1979, a group of residents had formed a group known as the Concerned Citizens Against the Fire. When that effort proved futile, the name was changed to the Central Committee for Human Development and the goal was to obtain federal funds for the purchase of the Centralia houses and to allow the residents to relocate elsewhere.

One of the leaders of the group was Joan Girolami. As with Lois Gibbs at Love Canal and Judy Piatt at Times Beach, she had had no previous experience with citizens' action groups. As with the other two, she was a homemaker with two children. Also as with the other two, she was prompted to action by a specific incident built on a longer history of mistrust.

> I think it was a slow buildup of government lies and hot temperatures that got me. In the beginning I was

apathetic, like everyone else. In 1969 they drilled holes in my yard and told me there was no fire there. So I forgot about it.

We wanted to put in a pool, so we went to the Bureau of Mines and asked if the fire was there. And they said, "No, no, no, go ahead and put the pool in." So we spent a couple thousand dollars and put a pool in. Then they put in more boreholes and found the temperature in my backyard was seven hundred degrees. I was furious, and I had never gotten anything in writing from them when they told me it was safe.

Not until 1982 did the Office of Surface Mining acknowledge that the mine fire had breached the fly ash barrier, and in 1983 they declared that the fire was out of control.

Indeed, Lois Gibbs, now a celebrity in the environmental movement, talked to the people of Centralia and the attendant media in 1983. Her words are worth repeating. They provide an insight into what is involved and what forces such citizens' groups sometimes have to face. To be sure, her words do portray government, industry and science in an overly black light with which, as a generality, we disagree; but they do correctly describe her experiences and that of other similar groups.

> The only thing that resolves problems of this sort is people banding together. When you look at Love Canal, it wasn't the fact that 56 percent of the babies were born birth-defected that initiated our evacuation. It was an election campaign. That's the way the system works. The media is the only way to move public officials to do what's right, and the media reacts to mothers and children. Politicians know that. The only way to stop companies is economic. Dow, Hooker, or Union Carbide: If they don't pay, they're going to do what they're doing. They don't have morals, they don't have principles. They only know profit. Once you affect their profit, then they'll change.

In Times Beach it had absolutely nothing to do with the fact that they were exposed to dioxin. That evacuation came solely because Anne Gorsuch [Burford] and Rita Lavelle's heads were on the chopping block. Love Canal, Times Beach, and now Centralia are all major public embarrassments.

The people who are fighting with political action, with voices, with togetherness and common sense are winning. The first thing that people think they need to do is hire a lawyer. The second thing is to get a hydrologist or toxicologist. That is such a wrong move. They don't need to hire experts. It wastes money, and people have to realize that these are not scientific issues, but political issues. Nobody knows what the mine fire gas is going to do to folks over a twenty-year period. Or dioxin. You can't argue it scientifically; therefore, you can't argue it legally. There is no such thing as objective science when politics is involved.

Even though we were winging it in Love Canal and learning from the seat of our pants, we did well because we knew what we were doing politically. The hardest thing to convince people of is that their families' dying from toxic exposure, be it from mine gas or dioxin, is not going to get them out. And women will say, "No, no, no, I can't organize. I'm not a Lois Gibbs." And I'll say, "Of course you are." What's a Lois Gibbs. A Lois Gibbs is a housewife. Any housewife in America knows how to organize, budget, and delegate responsibility. And women tend to be highly involved in grass-roots organizing. Disasters seem to hit low- to moderate-income, minority, bluecollar communities where the men are working all the time on swing shifts. The country club communities usually escape. And the government plays on their feelings of powerlessness by telling the people that they're not experts, they've never gone to college, and they're just dumb house-

Lois Gibbs in Centralia. From Jacobs, 1986. Renée E. Jacobs

wives. If the bureaucracy would give people straight information, the amount of emotional stress would be much less. Instead they say, "We have these Ph.D.'s, this consulting firm, and we're going to do this health study and that survey," and people just feel like guinea pigs.

But there's no way the "scientists" are objective with all these dollar signs looking them in the face. You have to realize the bureaucrats initiating the studies are also going to be paying for the resolution to the problem. The same is true of people from universities doing studies, because their funding generally comes from the state or corporations. They're going to do a twenty-million-dollar study on Love Canal. For what? They're trying to revitalize it and sell the homes.

Thanks to a donation of $30,000 from the Catholic Church and the able assistance of Sister Honor Murphy, the Central Committee continued its efforts. By 1984 the federal government had approved an appropriation of $42 million to buy out the houses of those residents of Centralia who chose to take advantage of the offer. Many, but not all, did so.

It is ironic to look back on this environmental disaster which cost the federal and state governments a total of $49 million—$42 million for the buy-out and $7 million for the unsuccessful mine fire control—when an extremely modest expenditure of $175 at the outset might have prevented the whole thing.

The fire still burns.

CHAPTER 5

On a Regional Scale

London and Los Angeles—Harold Des Voeux

We generally associate pollution with our own modern era. Yet even before the industrial age, environmental problems called for protective measures. Smoke from wood fires in primitive abodes with poor ventilation has probably always been an annoying but acceptable irritation; but smoke from bituminous (soft) coal, introduced in England in the 1200s as an inexpensive substitute for wood to heat shops and homes, was even more annoying. It combusts very inefficiently, creating great quantities of airborne soot and noxious sulfur dioxide. It was tolerated, however, for simple economic reasons.

But in the 1300s, Kind Edward I issued an edict prohibiting London merchants from using coal in their furnaces during sessions of Parliament, ordering a return to the practice of burning wood. The penalties were severe: one man, upon being caught with warm coals in his furnace, was put to death. This is the first documented piece of environmental legislation. (It was around this time, however, that the Hopi Indians in Arizona found that they could burn coal and did so for making pottery. Legend has it that the smoke smelled so bad that the tribal leaders ordered a kachina—a masked figure representing a spirit-being—to go to all the homes and ban the use of coal indoors.)

In due course, the London merchants prevailed, and the king's edict failed. Episodes of smog continued to plague the city. These were the result of coal smoke and fog, for which London became famous and from which the word "smog" was later coined. Through the years, people continued to object. One such was John Evelyn, who wrote a pamphlet in 1661 with the elaborate title, *Fumifugium, or, the inconvenience of the air, and smoke of London dissipated. Together with some remedies humbly proposed by J. E. Esq., to his Sacred majesty, and to the Parliament now assembled*. The pamphlet begins:

> Sir, it was one day, as I was walking in Your Majesty's Palace at Whitehall (where I have sometimes the honor to refresh myself with the sight of Your Illustrious Presence, which is the joy of Your People's hearts) that a presumptuous smoke issuing from near Northumberland house, and not far from Scotland yard, did so invade the Court, that all the rooms, galleries and places about it, were filled and infested with it, and that to such a degree as men could hardly discern one another from the cloud, and none could support, without manifest inconvenience. It was this alone and the trouble that it needs must give to your Health, which kindled this indignation of mine. Nor must I forget that Illustrious and Divine Princess, Your Majesty's only Sister, the now Duchess of Orleans, who, late being in this city, did in my hearing, complain of the effects of this smoke both in her breast and lungs, whilst she was in Your Majesty's Palace.

Evelyn continued unsuccessfully to seek smoke abatement in London. In 1684, he wrote a note of interest to the scientific historian, being perhaps the first, albeit indirect, notice of a temperature inversion.

> London, by reasons of the excessive coldness of the air hindering the ascent of the smoke, was so filled with the fuliginous steam of the sea [bituminous] coal, that one could hardly see across the streets, and this filling the lungs with its gross particles, exceedingly obstructed the breast, so as one could hardly breathe.

No amount of complaining would stop the use of coal, and in the 1800s the people of London seemed to take a perverse pride in their smogs, as the use of soft coal increased for both residential and industrial purposes. They began referring to them as "pea soupers", from the yellow color of the smog, or as "London particulars", a word coined by Charles Dickens in his novel *Bleak House*. The heroine describes how a stagecoach driver introduced her to London:

> He was very obliging; and as he handed me into a fly, after superintending the removal of my boxes, I asked him whether there was a great fire anywhere? For the streets were so full of dense brown smoke that scarcely anything was to be seen.
> "O dear no miss," he said. This is a London particular."
> I had never heard of such a thing.
> "A fog miss," said the young gentleman.

In his story *The Bruce-Partington Plans* Arthur Conan Doyle vividly describes, via Watson, a proper pea souper:

> In the third week of November, in the year 1895, a dense yellow fog settled down upon London. From the Monday to the Thursday I doubt whether it was ever possible from our windows in Baker Street to see the loom of the opposite houses. The first day Holmes had spent in cross indexing his huge book of references. The second and third had been patiently occupied upon a subject which he had recently made his hobby—the music of the Middle Ages. But when, for the fourth time, after pushing

back our chairs from breakfast we saw the greasy, heavy brown swirl still drifting past us and condensing in oily drops upon the window panes, my comrad's impatient and active nature could endure this drab existence no longer. He paced restlessly about our sitting room in a fever of suppressed energy, biting his nails, tapping the furniture, and chafing against inaction.

"Nothing of interest in the paper, Watson?" he said.

I was aware that by anything of interest, Holmes meant anything of criminal interest. There was news of a revolution, of a possible war, of an impending change of government; but these did not come within the horizon of my companion. I could see nothing recorded in the shape of crime which was not commonplace and futile. Holmes groaned and resumed his restless meanderings.

"The London criminal certainly is a dull fellow," he said in the querulous voice of a sportsman whose game has failed him. "Look out the window, Watson. See how the figures loom up, are dimly seen, and then blend once more into the cloud bank. The thief or the murderer could roam London on such a day as the tiger does the jungle, unseen until he pounces, and then evident only to his victim."

"There have," said I, "been numerous petty thefts."

Holmes snorted his contempt.

"This great and somber stage is set for something more worthy than that," said he. "It is fortunate for this community that I am not a criminal."

"It is indeed!" said I heartily.

In a speech in 1905, a British physician named Harold Des Voeux introduced the term *smog*, to describe the smoke and fog conditions that affected London. In 1929 he joined with others to form the National Smoke Abatement Society. Five years later, as president of the Society, he reported on their progress—or rather lack of progress—in generating interest in the London smog problem.

> Surely, we have preached and preached until our brains are empty…and how many disciples have we procured? Even in this enthusiastic city how many of its million inhabitants care two straws whether the air is clean or dirty…? Where is the fault? Is it ours or theirs?
>
> We do not want to create alarm but we do want those who live in our cities to recognize that they live under unhealthy conditions conducive to maladies which not only cause discomfort and annoyance and temporary disablement, but dire diseases of a nature persistent enough finally to undermine strength and vitality and to lead to an untimely end. What will arouse the man in the street? For unless we can do this the fight will continue for another thirty years.

The unfortunate answer to Des Voeux's questions was that it would take a crisis and a tragedy with a sufficient number of victims to raise the interest of the public and the government to the cause of clean air. It is a theme that has been played out in much the same fashion many times since—little foresight and much hindsight.

There were small incidents which received only passing attention. In December 1930 thirty people in Belgium's Meuse Valley died from smog, and thousands more were affected. Seventeen people died from smog in Donora, Pennsylvania in October 1948; and six thousand people were affected, representing a little less than half the population. In both cases, smog enveloped the narrow valleys of these two highly industrialized regions and took the lives of elderly people and those with chronic lung and heart diseases.

From such incidents, it gradually came to be appreciated that a major noxious component of smog was sulfur dioxide. In the humid fog, the sulfur dioxide was converted to sulfuric acid,

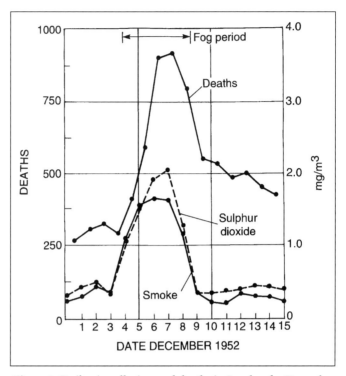

Figure 8. Daily air pollution and deaths in London for December 1952. From Meetham et al., 1981.

which in turn was adsorbed onto the fine airborne particles of soot. The membranes lining the eyes, nose, and especially the respiratory track are most vulnerable to these toxic particles; and the results can be bronchitis, bronchial asthma, emphysema, and lung cancer. Beyond human damage, smog blackens buildings with soot and the sulfuric acid erodes both buildings and statuary.

But the crisis came in 1952. On Thursday, December 4, a large high pressure system arrived and sat over London, gradually lifting five days after it arrived. All the smoke from the millions of residential and industrial chimneys simply accumulated day by day in the cold and motionless foggy air. It was the most intense and long lasting smog on record, the "killer smog" of 1952. Four thousand people died, chiefly the elderly and those with chronic respiratory problems. The accompanying graph shows the smoke and sulfur dioxide levels for the killer smog and the number of deaths each day in London during this period. Many thousands more may have been permanently affected; an estimated 50,000 to 100,000 were made sick at the time.

The black and yellow smog was so thick that you couldn't see a foot in front of you. People who ventured out wearing white returned with gray clothing; those with colds coughed up black mucus. People got lost in their own neighborhoods. One doctor lost his way trying to visit a patient and wound up on his own doorstep; not to be undone, he enlisted a blind neighbor who led him to the patient's house. Traffic came to a standstill, or barely moved. Cars were abandoned and, in one instance, a policeman on foot, leading a convoy of cars out of London to a nearby suburb, was overcome by smog after walking two miles. Elsewhere, a driver ended up in a cemetery, without knocking down any tombstones and without the slightest idea of how he got there.

After several hundred years of inaction, the killer smog impelled the government to do something about air pollution. In 1956, Parliament enacted the Clean Air Bill, prohibiting the burning of soft coal in London, and the days of pea soupers were soon over. There is clean air over London, and much of the statuary and building facades have been cleaned of their black and sooty coating.

Across the world, and a little earlier than the killer smog of London, in the early 1940s, the people of Los Angeles began to notice a light blue haze that would occasionally settle over the city, sometimes remaining for days at a time. With it came reduced visibility and varying degrees of eye, nose, and throat irritation. At first authorities decided the irritant was the same as that found in London smogs—sulfur dioxide—and sought to reduce emissions from various industrial sources. But this produced no noticeable result. Further laboratory investigations showed that petroleum vapors, in combination with photochemical reactions from the Sun's ultraviolet radiation, produced the same compounds observed in the Los Angeles haze. Again, authorities sought to reduce the escape of petroleum vapors from local storage tanks and oil refineries, and again the restrictions didn't help.

A vast network of buses and trolleys that linked the various neighborhoods of greater Los Angeles had begun to be dismantled in the previous decade and the famous freeways were being built. The automobile had become virtually the only means of transportation, and it was finally appreciated that the stinging haze resulted from millions of cars releasing thousands of tons of hydrocarbons into the air daily, thanks to the inefficiency of the combustion system.

The Los Angeles pattern has been repeated in many other environmental case histories. It is comforting to be able to point a finger at a specific entity, like an industry, as the polluter. Corrective action can be taken without much harm except to that entity. But it is extremely uncomfortable—and harder to correct—when it turns out, as Pogo said, "we have met the enemy and it is us."

The photochemical haze (sometimes inaccurately referred to as smog) that haunts Los Angeles is made worse by two meteorological phenomena. First, the surrounding mountains essentially prevent air circulation within the Los Angeles basin; the only movement of air is a gentle breeze from the Pacific Ocean. The second phenomenon is the temperature inversion. Generally, temperature decreases upward from the Earth's surface, but not always. Often, warm dry air from the desert to the east flows into the Los Angeles basin at elevations of 1,500 to 3,000 feet, effectively putting a lid over the area and preventing hydrocarbon emissions from escaping upward. Instead, they spread out over the city in a blanket of noxious haze. There are few cities that do not have occasional temperature inversions, and fewer still that do not suffer from photochemical haze. But the Los Angeles basin, which includes all of Orange County and the non-desert portions of three others, still has the worst air pollution in the United States, failing to meet federal air quality standards for four of the six criteria pollutants. It meets the standards for lead and sulfur dioxide, but it is the only area in the nation that fails to meet the standard for nitrogen dioxide; ozone levels sometimes reach three times the health standard, and carbon monoxide and fine particular matter are often twice the legal limit. Two days out of three, the air is judged to be unhealthy. How unhealthy? It is estimated that if the area met the federal standards, medical costs would be reduced by $9.4 billion a year.

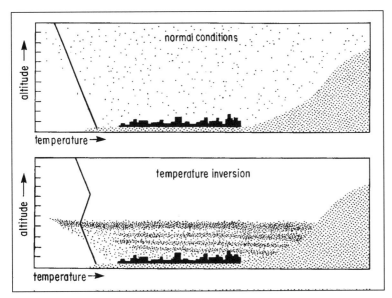

Figure 9. On a clear day in Los Angeles, the temperature decreases with increasing altitude. When the area is capped by a mass of warm, stable air, the temperature increases for several thousand feet, preventing the dissipation of pollutants from below. Adapted from Wagner, 1971.

At the same time, nowhere in the United States are more innovative, even extreme, efforts being made to cure the problem. A three-tiered program has been adopted by the area: (1) In the early 1990s, tighten tailpipe standards, use buses that burn "clean" fuels like methanol, ban consumer products that contain haze-producing chemicals, phase out coal and fuel oil in industrial plants and power plants; (2) by about 2005, forty percent of cars, seventy percent of trucks, and all buses will run on "clean" fuels, and there will be reduced emissions from such vehicles as aircraft, ships, and locomotives; and (3) the final stage requires the use of technologies that are not fully developed—the use of fuel cells and such fuels as hydrogen, and the conversion to electricity production without fossil fuels, relying instead on solar, wind, and geothermal energy.

Photochemical haze appeared first in Los Angeles, ever the trendsetter. Now, of course, it occurs in New York, in Houston (situated on a flat coastal plain with no mountains), in the mile-high city of Denver, where one might expect pure mountain air, in Phoenix, where people used to go to find relief from asthma—the list is long. And it may well be that most cities will have to follow the lead of Los Angeles and, as the English did after the killer smog, get serious.

There has been remarkable progress over the past two decades in cleaning up some forms of air pollution, particularly in the United States, Canada, and much of Western Europe. But much of the world's urban population still breathes air that is unhealthy. What usually begins as a local problem in many cases balloons into a regional problem and, sometimes, as we will see later, even a global one.

Lake Erie and Chesapeake Bay—Chesapeake Bay Foundation

Water pollution like air pollution is a common problem. And like air pollution it is related to the excessive waste products from our ever burgeoning human population.

Locally, water pollution is a common phenomenon. Discolored water and unsightly rafts of algal blooms or aquatic plants are common in small ponds in both urban and rural areas. Lake beaches are often closed to swimming for health reasons, and freshwater supplies have been reduced. Many coastal bays around the country are closed to shellfishing from time to time because of pollution.

Here, we look at two large and prominent examples of water pollution. Chesapeake Bay, the largest estuary in the United States, and Lake Erie, the fourth largest and the most polluted of the Great Lakes.

The basic biological cycle is much the same for both lakes and estuaries. We start with the anthropogenic inputs. First, there are the organic wastes from humans and animals. These are oxidized through bacterial action into inert inorganic compounds. This is accomplished either in the receiving water body or in a sewage treatment plant or septic tank prior to discharge or runoff to the receiving water body. The size of the organic pool discharged to the lake or estuary is expressed in terms of how much oxygen is required for the process—its biochemical oxygen demand, or BOD (see Figure 10).

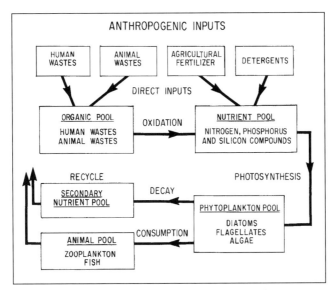

Figure 10. Basic biological cycle in lakes and estuaries. A critical portion of the cycle is the phytoplankton pool. Increased nutrient availability can lead to unsightly plankton blooms and to anoxic conditions as the available oxygen is used up in the plankton decay. From Officer and Page, 1993.

The non-organic products that result from the oxidation are referred to as *nutrients*. They include nitrogen in the form of nitrate, nitrite, and ammonia and phosphorus in the form of phosphate. A third nutrient, silicon in the form of silicates, enters the system through the weathering of soils and rocks. In addition, there are the direct additions of nutrients from human activities in the form of excess fertilizer runoff from agricultural lands and detergent discharges from municipal regions.

The nutrients, along with carbon dioxide, feed or nourish—hence, the term nutrient—the process of *photosynthesis*. Photosynthesis is the beginning step in the biological cycle and in the aquatic realm leads to the creation of phytoplankton. An amazing process, living matter from non-living sources. Phytoplankton are the microscopic plants that drift around in the water—diatoms, which need silicon for their shells; flagellates; and green and blue-green algae among them. Some of these nutrients may also go into producing a complementary pool of rooted aquatic plants.

Under normal circumstances, the phytoplankton pool is consumed by zooplankton, i.e., microscopic animals, and herbivorous fish. And these are, in turn, consumed by carnivorous fish. However, when the nutrient pool is overloaded by various human inputs—nitrogen and phosphorus from fertilizer runoff (used to stimulate crop photosynthesis), for example, or phosphorus from detergents—the phytoplankton pool changes in both quantity and quality. First, with the addition of vast quantities of nutrients, the population of the phytoplankton grows enormously—far more than the zooplankton can eat. And because the level of silicon drops compared to the great increase in nitrogen and phosphorus, diatoms become relatively scarce. The diatoms, which are prime quality food for the zooplankton, are replaced by flagellates, a poorer food source, and by green and blue-green algae, which have no nutritive value whatever. Thus, most of the teeming additions to the phytoplankton pool form a dead end. They, themselves, must be depleted by decay, not by useful consumption. The result of such excessive loading of the aquatic environment—with its excessive plankton pool and oxygen demand for the decay of the plankton pool—is called *cultural eutrophication.*

One sign of eutrophication is the unsightly algal blooms at the surface of affected lakes or bays. A more lethal sign often lies beneath the surface. The unused planktonic materials—chiefly algae and flagellates—eventually sink to the bottom and decay very slowly, over months, using up the available oxygen in the water. This presents no special problem in the winter, when the bottom waters of lakes and estuaries are usually replenished with oxygen from atmospheric influxes at the surface. But in spring trouble starts. The surface waters of lakes are heated and a *thermocline* develops between the oxygen-rich surface waters and the waters of the bottom. Literally a heat boundary, the thermocline, is a barrier that prevents surface and bottom waters from mixing. A similar process occurs in estuaries, when the spring runoff of fresh riverine water adjoins and overrides the more saline and denser waters of the estuary's bottom. In this case a salt boundary—a *halocline*—forms, serving as a barrier to the mixing of surface and bottom waters. In both cases, however, the planktonic debris keeps on decaying, using up the oxygen in the bottom

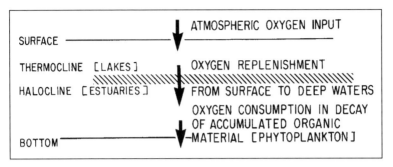

Figure 11. Oxygen dynamics for lakes and estuaries. When the rate of oxygen consumption in the decay of the accumulated phytoplankton material on the bottom is greater than the net oxygen replenishment from the surface waters, the oxygen levels for the deep waters will decrease and eventually lead to an anoxic condition. From Officer and Page, 1993.

waters until a situation of low oxygen is reached, called *hypoxia*. Or worse, *anoxia*, which means no oxygen at all. Creatures that require oxygen to sustain life naturally cannot live in such places; the region is dead. The death of these waters persists through the spring, summer, and fall months until, with the onset of winter, the surface and bottom waters begin to mix again (see Figure 11).

This is what has happened to the nation's largest estuary, the Chesapeake Bay, which drains a vast portion of the Middle Atlantic states, including parts of Pennsylvania, Delaware, Maryland, and Virginia as well as Washington, D.C. The overall bay system includes a number of tributary estuaries, notably the Patuxent, Potomac, Rappahannock, York, and James. And its principal source of freshwater is the Susquehanna River. The amount of municipal and industrial waste, along with agricultural runoff, that pours into the bay is enormous. In 1971, 69% of the phosphorus put into the bay was from municipal and industrial wastes; and 66% of the nitrogen—the more important of the two nutrients, as it turns out—came from agriculture. But the excessive loading process has been going on for a long time and continues at increased loading levels to today.

As far back as the 1930s, it was observed that the deep waters of the midportion of the bay became anoxic for short periods in the summer. During the entire summer of 1950, all the deep

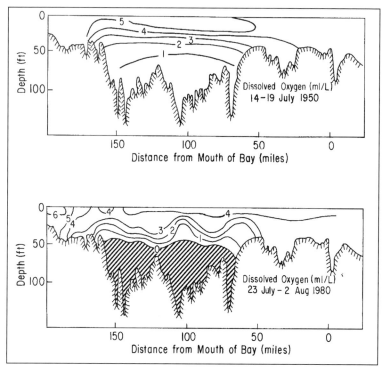

Figure 12. Dissolved oxygen levels in milliliters of oxygen per liter of water along the main channel of Chesapeake Bay extending from the Susquehanna River to the Atlantic Ocean during the summers of 1950 and 1980. Diagonal region is devoid of oxygen. From Officer et al., 1984.

waters of the bay channel had dissolved oxygen levels considered hypoxic—that is, 1-2 milliliters of oxygen per liter of water. By the summer of 1980, these same waters had become utterly anoxic, beginning in late spring and continuing until early fall. And hypoxic conditions extended over most of the rest of the bay bottom, extending as well up into the Patuxent and Potomac estuaries (see Figure 12).

Not surprising, there have been major changes in the populations of plankton, shellfish, and fin fish—and substantial ecological and economic disruption.

In the earlier decades of this century, the phytoplankton in the middle and upper regions of the bay were dominated by a prominent spring bloom of diatoms, along with a smaller winter bloom.

By the 1980s, these same waters experienced a single large and extended bloom that peaked in summer, dominated by flagellates and green algae.

There has been a devastating effect on the bay's blue crabs, a prominent, almost legendary, seafood staple for the Chesapeake region. (Maryland crab cakes are renowned throughout the country.) By the early 1950s, commercial crabbers reported the death of crabs in the midbay region. Field observations at the time showed that there was greater than 50% mortality when oxygen levels reached the hypoxic state. Before 1965, crab fishermen from midbay Tilghman Island found an abundance of crabs in the deeper waters both early in the season, prior to mid-May, and after mid-September. By 1983, there was no more deep-water crabbing. In earlier years, the crabs had hibernated in the mud under the protected deeper waters; by 1983 they were hibernating in shallow waters. Indeed, in 1982, no crabs were caught in waters deeper than twelve feet; and many that were caught were in a poor condition, dying before they could be brought to market. Things have not improved in the ensuing years.

In the midbay area, "crab wars" have often occurred, in which tens of thousands of crabs crowd into the shallows and even sometimes crawl out onto the land. This bizarre behavior is probably the result of a seiche. A seiche occurs when the wind blows steadily across the bay. The surface waters tend to pile up on the windward shore, and there is a compensatory flow of deeper and poorly oxygenated bottom waters toward the lee shore. This later flow obliges the crabs to flee into very shallow depths. Crab wars have also been reported for the Potomac estuary, where, in 1973, all crabs below a depth of eighteen feet died in the hypoxic water.

Oysters, also a Chesapeake Bay seafood staple, have fared equally poorly. Virginia oystermen have reported "black bottoms" in the later summer of recent years. There dead oysters and other creatures of the bottom are found in black, foul-smelling, anoxic sediments.

Hypoxia and anoxia are, of course, no better for fin fish than for shellfish. Here too there has been a serious economic decline and a large ecological shift. The Chesapeake Bay had long been a major source of fish like the alewife, American shad, striped bass, and white perch—anadromous fish that travel up the estuary in the spring to spawn in shoal waters, with the juveniles heading down the estuary later in the season. From 1960 to 1975, the bay contributed 3,000 tons a year of striped bass alone to American tables; by 1980, the catch was about 700 tons.

Yet other marine spawners, like spot and croaker, tend to enter the bay in late spring and summer in the form of larvae and juveniles. They are bottom feeders, and anoxia of the bottom waters from May to September restricts their habitat and food sources. Commercial landings of croaker dropped from 20,000 tons annually before 1950 to about 5,000 in succeeding years.

Meanwhile, a less desirable fish has been on the upswing. Landings of menhaden have steadily increased over the past thirty years. These are herbivorous fish that can stand substantial environmental stress (such as low oxygen levels) and they are commonly thought of as "trash fish". They are used chiefly for fertilizer and animal feed. Menhaden now account for 90% of the total fin fish landings in the Chesapeake Bay, a sorry result for a once-rich fishing ground.

Much the same degradation has occurred for Lake Erie. But something has been done there to alleviate the eutrophication, which has not been the case for Chesapeake Bay.

Lake Erie is the shallowest of the Great Lakes, with an average depth of sixty feet, so it simply doesn't have the assimilative capacity of the others. Furthermore, it has been calculated that it would take two to three years to drain Lake Erie through its natural outlet to Lake Ontario. Not that anyone has suggested doing so; the calculation provides us with a residence time of a particle of water in the lake, the average time that any given particle remains in the lake. Two to three years is a long period of residence, and it means that noxious components—an overabundance of nutrients and toxic wastes—that seep or are dumped into the lake stay in the lake. Beyond that, the shores of Lake Erie are lined with more industries and more cities and towns than any of the other Great Lakes. All of the human and industrial wastes from Buffalo, Erie, Cleveland, Toledo, and even Detroit (via the Detroit River) wind up in the lake along with fertilizers, herbicides, and pesticides running off from the region's farmlands.

As with Chesapeake Bay, an early sign that the lake was in

trouble came from the fishing industry. In the 1920s, landings of cisco, whitefish, pike, and sturgeon were around fifty million pounds a year. By 1965, landings of these commercially valuable fish had collapsed to a mere 1,000 pounds. The drop in sturgeon could be attributed to overfishing, but not the rest. As in the Chesapeake Bay, there has been a weedy substitute. Fish landings remain near fifty million pounds, but they are what the locals call "rough fish": catfish, carp, and other less palatable types. These less valuable fish are warm-water creatures with lower oxygen level requirements, while cisco, pike and whitefish are cold-water fishes. They dwell near the bottom and, as the deep waters of Lake Erie headed for anoxia, these fish headed for oblivion.

By the late 1960s and early 1970s, it was evident that the deep waters of the central basin were going anoxic in the summer. Excessive and obnoxious algal blooms would from time to time cover large portions of the lake's surface. Swimming had to be prohibited in some areas because of untreated sewage and mats of decaying vegetation littering the beaches. Toxic wastes and oil scum haunted the harbors and their entrances. Action was necessary, and over time it was taken. Sewage treatment plants were built; some toxic wastes were curtailed. And most important, there is now a ban on the use of detergents containing phosphorus. From 1940 to 1970, the phosphorus loading to the lake increased three-fold, nearly all of it from detergents. A ban on the use of phosphorus in detergents was imposed in 1970. Since then, the total phosphorus input to the lake has decreased by a half.

A reasonable question: why concentrate on phosphorus over the other principal nutrient, nitrogen? In photosynthesis, nitrogen and phosphorus combine in an atomic ratio of sixteen to one. Photosynthesis will therefore be limited by whichever of the two nutrients is in the more limited supply according to the combining ratio of sixteen to one. In the oceans, nitrogen is usually the limiting nutrient; but with the great overloading of nitrogenous waste into Lake Erie, phosphorus was the limiting nutrient. Even major curtailment of the supply of nitrogen would have little to no effect on the algal blooms. But by limiting phosphorus, the photosynthetic process itself is limited, thus cutting back on the plankton crop. So the degradation of Lake Erie has been interrupted by the ban on detergents containing phosphorus, but the lake is a long way from being in the nearly pristine state it enjoyed before the industrialization of its shores.

For Chesapeake Bay, nitrogen is the limiting nutrient. That is a much more difficult problem to deal with than banning phosphate detergents for which an industrial substitute could be provided. As mentioned, 66% of the nitrogen loading to Chesapeake Bay comes from agricultural runoff. There are, essentially, no curbs on the amount of fertilizer that a farmer can use. And it is generally more economic to use an excessive amount of fertilizer with consequent increased crop yield but also increased unwanted runoff than to curtail the amount of nutrient input to the soil to that taken out in crops.

But there is hope for Chesapeake Bay. Once again through the actions of individuals—in this case, in the form of citizens' groups. For the bay, as a whole, these groups include the Chesapeake Bay Foundation, Alliance for Chesapeake Bay, and Chesapeake Bay Trust. In addition, there are several other smaller groups with specific interests in individual tributary estuaries, rivers, streams, and associated wetlands and land trusts; in Maryland, alone, there are over seventy such organizations. The oldest of the groups is the Chesapeake Bay Foundation with their motto, Save the Bay. The foundation was started in 1966 and now has a membership of over 80,000 and an annual budget of around $8,000,000. with the funding coming mainly from membership contributions and grants and gifts. The activities of these various groups include environmental education; increasing land trusts; influencing environmental legislation; pursuing through legal means polluters who are not in compliance with the law; reducing fertilizer, sewage, and toxic inputs to the bay; conserving fisheries; expanding wetlands and providing habitat protection; transforming agricultural practices; and influencing land use and transportation. A large order but not an impossible one.

And the efforts in Chesapeake Bay have been replicated elsewhere in nearly every other estuary or bay throughout the country. These include, but are by no means limited to the following: People for Puget Sound, Save San Francisco Bay Association, Save the Bay (Narragansett Bay), Save the Sound (Long Island Sound),

Tampa Baywatch, and Galveston Bay Foundation. There is great concern for protecting our estuaries, bays, and associated wetlands. There is a concerned and informed public out there and they are active.

Acid Rain—Noye Johnson

As with the air pollution over London and Los Angeles and the water pollution of Lake Erie and Chesapeake Bay, the acid rain problem has been recognized for some time. And a great deal has been done to alleviate its effects.

As long ago as the 1960s, a group of scientists then at Dartmouth College—Herbert Bormann, Gene Likens, Robert Reynolds, and the late Noye Johnson—were concerned by the effects on New England forests from what they were certain was acidic rain produced from the burning of fossil fuels, specifically the sulfur dioxide emissions. At the same time, fishermen in Scandinavia, then in Scotland and North America, began to notice a striking decline in both the quantity and quality of fish that were caught in some remote lakes. It was not long before the same lakes were shown to have become more acidic.

The measure of the acidity (or conversely, alkalinity) of a solution is the pH number. The pH scale ranges from 0 to 14, with a neutral solution having a value of 7. Anything greater than 7 is alkaline (or basic); anything less than 7 is acidic. The pH scale is logarithmic so a solution with a pH of 4 is ten times more acidic than one that is pH 5, and 100 times more acidic than one that is pH 6.

By way of benchmarks, lemon juice has a pH ranging from 2.2 to 2.4. Vinegar is 2.4 to 3.4, and grape juice is 3.5 to 4.5. Normal rain, which is slightly acidic with a pH of 5.6, is at the same time ten times less so than grape juice. The moderate acidity of normal rain is a result of the combination of ambient carbon dioxide in the atmosphere with water vapor, forming minute droplets of carbonic acid, a relatively mild acid. Any rain with a pH of less than 5.6 is considered acid rain. Today the acid rain that falls on the northeastern part of the United States has pH values ranging from 4.1 to 4.3. In the summer clouds overhead, the lower portions have values around 3.6, sometimes reaching the acidity of vinegar (see Figure 13).

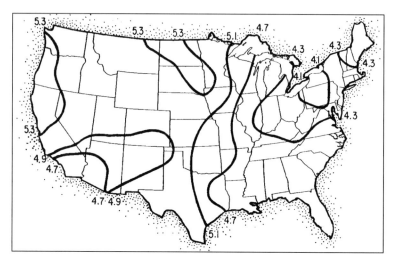

Figure 13. Acid rain precipitation in the United States. The contour lines chart the average pH of the precipitation. From Mohnen, 1988.

The acid in acid rain is sulfuric acid. It is formed when sulfur dioxide combines with water vapor in the atmosphere. Most of the sulfur dioxide released into the atmosphere comes from burning coal. In the United States, 85% of the coal burned is to produce electricity, and the great proportion of coal-fired power plants exist in the industrialized Midwest.

As the automobile was the culprit in producing photochemical haze, the coal burning, electric power plant is the culprit in acid rain. A somewhat easier problem to deal with politically since it is related to a specific entity—the power plants—and not to the public at large.

Between 1937 and 1988 coal consumption in the United States doubled. Early on, high smokestacks were installed to eliminate locally the foul emissions of smoke and soot and, of course, these products were dispersed downwind. The sulfur dioxide too was swept downwind—to the east in the latitudes of the prevailing westerlies—to be deposited as acid rain. This has pitted New England against the Midwest, Canada against the United States, with similar problems in Europe and in almost every industrialized area that has neighbors downwind.

Just as hypoxia and anoxia cause ecological rearrangements in a lake or estuary, so does a change in acidity. The fish at the top of the food chain—called high-trophic level fish and typically considered the most desirable—are the most affected while fish further down the food chain—low trophic level fish—can handle considerably lower pH levels. One Canadian lake that was intensively studied over a period from 1961 to 1973 during which the pH levels changed drastically showed the following results. The first species to disappear was smallmouth bass, followed by walleye, turbot, and lake trout at pH levels of 5.8 to 5.2. Then, as the pH dropped to 4.7, northern pike, white sucker, brown bullhead, pumpkinseed sunfish, and rock bass all disappeared. Yellow perch and lake herring were still spawning in the lake as of the end of the study in 1973 with the pH at 4.7.

The acidification of lakes was one thing. In the early 1980s trees in the Black Forest of Germany and other European forests began to die at an alarming rate. By 1985, it appeared that more than half the trees in the Black Forest were damaged. People suggested climate changes, drought, disease, insects...but none of these could be shown to be the culprits. Similar problems were spotted in the forests of eastern North America, especially those at high elevations. Indeed, in the past twenty-five years, more than half the red spruce trees in New York's Adirondacks, the Green Mountains of Vermont, and the White Mountains of New Hampshire have died. Acid precipitation is taken up in the roots of trees and tends to reduce their tolerance for cold. Canadian studies showed that the most sensitive trees include Douglas fir, eastern white pine, white ash, and white birch. Those that show intermediate effects include balsam fir, basswood, red pine, and white elm. Those that are tolerant of acid rain include balsam poplar, red oak, and sugar maple. The conclusion was, and is, inescapable that acid rain has had devastating effects on lake fish and forests, here and abroad.

Meanwhile, Canadian environmentalists and other groups were pointing the finger at the Midwestern power plants and seeking redress. In the late 1970s, the U.S. Congress decided to institute a program to evaluate the situation, the National Acid Precipitation Assessment Program (NAPAP). Its mandate was to study the science of acid rain and come up with recommenda-

National Acid Precipitation Assessment Program final report was 6,000 pages long. From Roberts, 1991.

tions for an action program—for example, answers to such questions as what environmental benefits would accrue from a reduction of, say, 20 percent of sulfur dioxide emissions.

NAPAP got started in 1980. After several changes in leadership and direction, it issued its report in 1990, all 6,000 pages. A photograph shows a staff member standing beside the report, which is the same height as she is. Now, who is going to read such a report? It cost $500 million to produce, the work chiefly of various government agencies. As can be imagined, a vast amount of data was collected. Yet this data contributed little to the *science* of acid rain. Data collection, per se, can be a mindless operation and cannot be equated with science. In a scientific study, the results from one investigation may lead to an entirely different approach toward understanding the problem, calling for new and different data. This did not occur in the years of NAPAP. The masses of data collected may or may not eventually be of value to science.

Nor did NAPAP produce any suggestions for action, in spite of a gigantic computer program devised for that purpose. Any useful computer model has to rest on those scientific understandings and mathematical relations that are the basic input to the program. Otherwise it's the old story: "garbage in, garbage out." What did come out were four major conclusions: (1) acid rain had adversely affected aquatic life in about 10 percent of Eastern lakes and streams; (2) it had contributed to the decline of red spruce at high elevations; (3) it had contributed to the erosion of buildings and materials; and (4) along with related pollutants and especially fine sulfate particles, it had reduced visibility throughout the Northeast and in parts of the West. Every one of these conclusions could have been stated without benefit of this half-billion dollar study.

Nevertheless, progress was being made even while NAPAP dithered. Impelled by public concern and legislation from Congress, many of the utility companies in the Midwest substituted low-sulfur fuels for high-sulfur coal. (This had the ramifying effect of putting high-sulfur coal miners like those in West Virginia out of work, and also increasing strip mining in the West, where the coal is of a lower sulfur content.) In addition, scrubbers have been installed on many stacks. In scrubbers the flue gas effluents are channeled through a limestone, $CaCO_3$, slurry. The slurry reacts with the sulfur dioxide, SO_2, producing calcium sulfate, $CaSO_4$, which is subsequently collected and disposed, and releasing carbon dioxide, CO_2, to the atmosphere. Scrubbers can remove from 50% to 90% of sulfur dioxide emissions, but with some cost in plant efficiency. These costs are ultimately passed along to the consumer. In any event, as a result of scrubbers and low-sulfur fuels, sulfur dioxide emissions have been reduced by 35% over the past fifteen years while at the same time the use of coal has increased by 50%—a substantial and commendable first step.

Ogallala Aquifer—Groundwater Management Districts

Let us return one more time to the High Plains region and the Dust Bowl of the 1930s. Prosperity has, indeed, returned to the region even though its climactic conditions still put it in a marginal category for farming. How did this change come about? It

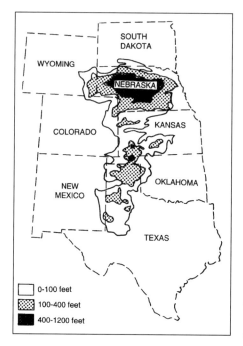

Figure 14. The High Plains (Ogallala) Aquifer. From Morgan et al., 1993.

relates strictly to the exploitation of an enormous source of underground water—the Ogallala Aquifer. As shown in the accompanying figure, the Ogallala extends in a north-south direction from South Dakota to Texas. To the north in Nebraska it has a thickness of over 400 feet; to the south in the Dust Bowl region it thins to around 100 feet or less.

The Ogallala is a fossil reservoir of water. It was formed principally during glacial times by runoff from the Rocky Mountains. It is the world's largest known aquifer and it presently supplies about 30% of our nation's irrigation water. The groundwater flow is from west to east at a slow rate of about 500 feet per year, and it continues to be replenished but only at a rate of one eighth that at which it is being depleted by pumping for agricultural purposes.

In other words, the Ogallala is being *mined*. It is not a renewable natural resource, as we normally think of water that is continuously cycled from rainfall, to plant and human consumption, to evaporation and runoff, and back to rainfall again. The Ogallala is a nonrenewable resource in the sense that it is being depleted at a more rapid rate than it is being replenished. Under present pumping conditions, it will go dry and the region will return to the barren High Plains that it once was unless corrective actions and changes in water consumption practices occur. For example, the Texas Agricultural Station at Lubbock—at the southern end of the Ogallala—has predicted that by the year 2015, irrigated land in that area will have to be reduced by 95%. And for the High Plains portion of Kansas the prediction is that 75% of the irrigated land will have returned to dryland acreage by 2020.

Serious mining of the Ogallala began only in the 1950s with the advent of high capacity pumps for lifting water from depths of 100 to 300 feet. As shown in the accompanying figure, the usual procedure is to feed the water to a center pivot irrigation system which consists of a long pipe, over a thousand feet in length, supported by A-frame towers. The A-frames move slowly, completing a circle in 12 hours or more and irrigate an area of 160 acres.

All the farmers in the region have known for some time that at high rates of underground water consumption, they face a big problem in the not-too-distant future—the end of the barrel. What they have done about it might appear to the skeptic as an impossibility. They have banded together for the *common* good. With appropriate enabling legislation of the late 1960s and early 1970s each of the Dust Bowl states, with the notable exception of Oklahoma, has formed a set of Groundwater Management Districts (GMDs) consisting of a few to several counties.

Each GMD is governed by a board of directors elected by the public either at a general election or at a widely publicized groundwater management district meeting. The boards hire a district manager or executive director and other staff as deemed necessary to carry out their policies. Their funding comes from a levy on the land owners using Ogallala water.

Their general objective may be considered to be conservation rather than strict preservation of Ogallala water. That is, the objective is to see that the water is most effectively used and in minimal quantities to curtail reservoir depletion. One may be assured, however, that the Ogallala will continue to be used. A few simple agricultural figures illustrate that. An irrigated cornfield

Center pivot irrigation system. From Zwingle, 1993. Jim Richardson, National Geographic Society.

produces 115 bushels an acre, compared with 89 bushels on an Eastern humid-land farm and 48 bushels on a dryland field.

The districts regulate well spacing, new well development, abandoned wells, and points of water diversion. They control groundwater that has been diverted or withdrawn from a source of supply and is not used, managed, or reapplied to a beneficial use; any act or omission causing the unreasonable deterioration of the quality of water; groundwater which an irrigator permits to escape or drain; and groundwater applied to a beneficial use in excess of the needs for such use. These are pretty far reaching controls, but the farmers, themselves, invoked them in an example of local populist management.

Answers to declining water supplies are also sought in more profitable alternative crops such as sunflowers and soybeans which are less demanding of water than the main staple crop of the region, corn. The GMDs also encourage better management practices like irrigation scheduling and more efficient technologies like drip irrigation. (Water use efficiency improves from 47% with furrow irrigation to 75% with scheduled sprinkling to 90% with drip irrigation.)

The GMDs actions have been effective. From 1950 to 1980 the aquifer's average water level decreased nearly ten feet, with declines exceeding 100 feet in some parts of Texas. From 1980 to 1990 the level declined only an additional foot with some parts of the system showing a rise in groundwater level. All of this was the result of increased rain and snow, better management practices, the use of new technologies, and a decrease in the land being irrigated. The land taken out of production during this period was 16 to 18% of the previously irrigated land and the principal reason for the decline was the increase in the price of fuel oil, fuel oil being the energy source for the pumps. Since 1990, 5 to 10% of the land has come back into production.

The goal of one of the districts in Kansas is to achieve zero depletion, i.e., the withdrawal of no more water from the aquifer than is being replaced by recharge. As one district official put it,

> the declining levels meant zero depletion anyway [at some time in the future], so why not opt to reach the same goal earlier while retaining an acceptable quantity of water for future management options.

All the districts will eventually have to take this as a goal if they want to see some form of profitable farming continue in the High Plains region.

CHAPTER 6

...and Even on a National to Global Scale

Pesticides—Rachel Carson

Every once in a while a giant comes along to bring public attention to a serious environmental problem and by a single act to put into motion corrective actions to alleviate the problem. In this case the giant was a soft spoken and diminutive woman named, Rachel Carson. And her action was the publication of a book, entitled, *Silent Spring*. As her editor, Paul Brooks, has stated it was "one of those rare books that have changed our thinking about man's place in the world of nature." And as her biographer, Robert Downs, has stated it was "comparable in its impact on public consciousness, and demand for instant action, to Tom Paine's *Common Sense*, Harriet Beecher Stowe's *Uncle Tom's Cabin*, and Upton Sinclair's *The Jungle*."

Rachel Carson was both a scientist and a writer. She was one of the few who could explain the wonders and discoveries of science to the public without in any way diminishing or simplifying the scientific content. In her several books she brought science and nature alive. Her writing at times reads like poetry. And the sales, measured in millions of copies, attest to her popularity.

Rachel Carson. From Brooks, 1987. Painting by Peter Darro for National Wildlife Magazine.

Rachel Carson was born in 1907 in Springdale, Pennsylvania. From an early age she knew that she was going to be a writer. Between the ages of ten and twelve she won three prizes and publication of her poems and stories. In 1925, she entered the Pennsylvania College for Women and graduated four years later having changed her major from English to biology. Thus began what would turn out to be her joint careers in science and writing.

In 1929, she attended the Marine Biological Laboratory in Woods Hole, Massachusetts, on a summer study fellowship and later that fall entered The Johns Hopkins University. She received her master's degree in marine zoology in 1932, having taught during that period and subsequently at both the University of Maryland and Johns Hopkins.

In 1935, she was able to land a job at the Bureau of Fisheries, a forerunner to the Fish and Wildlife Service, as a marine biologist. She was the second woman to be hired by the Bureau; women's rights were not high on the list of important things in the depression years of the 1930s with so many men unemployed. By 1948 she had risen through the Fish and Wildlife Service to be editor-in-chief of their publications. In December 1941 her first book, *Under the Sea Wind*, came out. This was just immediately prior to the bombing of Pearl Harbor and the United States entry into World War II. The book received little recognition beyond professional circles.

In 1951 things took an enormous upswing for Rachel Carson with the publication of her second book, *The Sea Around Us*, by Oxford University Press. Parts of the book were serialized by *The New Yorker* in June prior to book publication. (The serialization of the book had been rejected by over fifteen magazines before acceptance by *The New Yorker*, the list reading like a Who's Who of American magazines.) With that impetus, her excellent writing, and the complimentary reviews, the book became a best seller. In fact, it was on the list of best sellers for eighty-six weeks and was translated into thirty-three foreign languages. She was now famous. For the first time in her life she was relieved of financial worries. She left the Fish and Wildlife Service to concentrate on her writing.

In 1952 she received the prestigious National Book Award. In her brief acceptance speech, she spoke of her literary aims and philosophy of life:

> Writing a book has surprising consequences and the real education of the author perhaps begins on publication day. I, as author, did not know how people would react to a book about the ocean. I am still finding out.
>
> When I planned my book, I knew only that a fascination for the sea and a compelling sense of its mystery had been part of my own life from earliest childhood. So I wrote what I knew about it, and also what I thought and felt about it.

Many people have commented with surprise on the fact that a work of science should have a large popular sale. But this notion, that "science" is something that belongs in a separate compartment of its own, apart from everyday life, is one that I should like to challenge. We live in a scientific age; yet we assume that knowledge of science is the prerogative of only a small number of human beings, isolated and priestlike in their laboratories. This is not true. The materials of science are the materials of life itself. Science is part of the reality of living; it is the what, the how, and the why of everything in our experience. It is impossible to understand man without understanding his environment and the forces that have molded him physically and mentally.

The aim of science is to discover and illuminate the truth. And that, I take it, is the aim of literature, whether biography or history or fiction; it seems to me, then, that there can be no separate literature of science.

My own guiding purpose was to portray the subject of my sea profile with fidelity and understanding. All else was secondary. I did not stop to consider whether I was doing it scientifically or poetically; I was writing as the subject demanded.

The winds, the sea, and the moving tides are what they are. If there is wonder and beauty and majesty in them, science will discover these qualities. If they are not, science cannot create them. If there is poetry in my book about the sea, it is not because I deliberately put it there, but because no one could truthfully write about the sea and leave the poetry out.

There has also been a certain amount of surprise that a book in which human beings play little part should have been widely read. Like most authors, I have received many letters from readers, and perhaps you would be interested in the clues they offer. These letters started coming with publication of *The*

New Yorker profile, and they have never stopped. They come from all sorts of people, from college presidents to fishermen and from scientists to housewives. Most of the people say that it is because the book has taken them away from the stress and strain of human problems...that they have welcomed it.

They suggest that they have found refreshment and release from tension in the contemplation of millions and billions of years—in the long vistas of geologic time in which men had no part—in the realization that, despite our own utter dependence on the earth, this same earth and sea have no need of us.

"This sort of thing," wrote one reader, "helps one reduce so many of our man-made problems to their proper proportions."

Another said: "I am overwhelmed with a sense of the vastness of the sea, and properly humble of our own goings-on."

Such letters make me wonder if we have not too long been looking through the wrong end of the telescope. We have looked first at man with his vanities and greed, and at his problems of a day or a year; and then only, and from this biased point of view, we have looked outward at the earth and at the universe of which our earth is so minute a part. Yet these are the great realities, and against them we see our human problems in a new perspective. Perhaps if we reversed the telescope and looked at man down these long vistas, we should find less time and inclination to plan for our own destruction.

In 1955 she came out with her third book, *The Edge of the Sea*. As with *The Sea Around Us*, it was well received and was a best seller.

She then turned her attention to a problem that had concerned her while still at the Fish and Wildlife Service—the widespread and seemingly indiscriminate use of DDT and other insecticides and herbicides in agricultural and suburban control programs. Her concern, as well as that of many others, was that adequate consideration had not been given to the effects of these toxic substances on other fauna and flora besides the pests and weeds that they were intended to eliminate.

At first she tried to get someone else to write about the subject; but finally decided, if it were to be done, she would have to do it herself. With reasonable justification she felt that a book on such a technical and seemingly dreary subject—pesticides—would not have much public appeal. She also knew that she would be under attack from the chemical industries in her criticism of the callous manner that they treated these environmental poisons. However, with the encouragement of her editors at Houghton Mifflin and *The New Yorker*, Paul Brooks and William Shawn, she plowed ahead.

The result, *Silent Spring*, was over four years in the making. It was unlike her previous three books on science, nature, and the sea. It was a hard sell. The book came out in September 1962 having been serialized in three parts in *The New Yorker* the preceding June. With the serialization the book came under immediate attack from the chemical industries in the press, on radio, and on television, thereby indirectly adding enormous publicity for the book. *Silent Spring* became a best seller, in spite of Carson's reservations, comparable in sales to *The Sea Around Us* and greater in impact.

In light of the attacks that were made on the book as well as on Rachel Carson, herself, it is instructive to set out her thesis. In her words,

> It is not my contention that chemical insecticides must never be used. I do contend that we have put poisonous and biologically potent chemicals indiscriminately into the hands of persons largely or wholly ignorant of their potentials for harm. We have subjected enormous numbers of people to contact with these poisons, without their consent and often without their knowledge…I contend, furthermore, that we have allowed these chemicals to be

used with little or no advance investigation of their effects on soil, water, wildlife, and man himself…It is the public that is being asked to assume the risks that the insect controllers calculate. The public must decide whether it wishes to continue on the present road, and it can do so only when in full possession of the facts.

This is a most reasonable and well argued thesis. It is the responsibility of a given entity, be it industry or government, to be fully aware of all the effects of a potentially toxic substance before widespread use; an obligation to inform the public; and the ultimate decisions to be those of the public. *Silent Spring* went a long way to seeing that these practices were put into effect which, in our opinion, is generally the case today.

Silent Spring was a result of Carson's four years of study and research on the widespread and indiscriminate use of DDT as well as such other toxic compounds as chlordane, heptachlor, aldrin, endrin, dieldrin and malathion, which she collectively referred to as the "elixirs of death". It documents the effects of these insecticides and herbicides on other insects and plants and on birds, mammals, fish and, if somewhat speculatively, humans.

For Clear Lake, California, she documented the important process of *bioaccumulation*. It is the process whereby a toxic substance is increased in concentration from prey to predator, to next predator, and so on with, at times, disastrous effects on the highest members of the food chain.

From 1949 to 1957 the State of California applied selective doses of DDD, a close relative to DDT, to Clear Lake in maximum concentrations of 0.02 parts per million (ppm). The purpose was to rid the area of a small gnat that was a nuisance to fishermen and resort dwellers on the shore. The winter following the first spraying deleterious effects began to appear. Western grebes, fish eating birds, began to die with up to a hundred reported dead. By 1960 the nesting colonies of grebes had dwindled from 1,000 pairs to about 30 pairs and these few pairs did not produce any young. In 1959 the State of California stopped DDD applications to the lake.

Investigations during this period showed that the lake concentration of 0.02 ppm had increased to 5 ppm in the plankton, to 40-300 ppm in the herbivorous fish, to several hundred to a thousand ppm in the carnivorous fish, and to over a thousand ppm in the grebes. The grebes were the ones to suffer for a program aimed at exterminating gnats.

In another case, in 1958 the United States Department of Agriculture (USDA) embarked on a program to eradicate the fire ant in the Southern states. The chemicals used were dieldrin and heptachlor, both many times more toxic than DDT. The results were soon to come in. In Hardin County, Texas, virtually all the opossum, armadillo, and raccoon populations disappeared after the chemicals were laid down. Dead birds were prevalent in the treated areas. On a tract in Alabama which had been treated half of the birds were killed. On a 2,500 acre tract in Texas treated with heptachlor all the quail and ninety percent of the songbirds were lost. On a similar tract in Alabama a quail census had counted 121 birds prior to spraying; two weeks after treatment only dead quail were found. By 1960 with this kind of data coming in most of the State and County farm agents declined to participate further in the Federal fire ant eradication program and it gradually fizzled out.

In 1956 the USDA set out to eliminate the gypsy moth in the states of Pennsylvania, New Jersey, Michigan, and New York. Around a million acres were sprayed, this time with DDT. In 1957 a suit was brought by a group of citizens on Long Island to prevent the aerial spraying from extending to their region; their suit was denied and another three and a half million acres were sprayed. In suburbia, children at play and commuters at rail stations were showered with the oily mixture. Automobiles were spotted; flowers and shrubs were ruined; and birds, fish, and other insects were killed—but not the gypsy moth. Milk and vegetables were found to have DDT levels above the legal maximum for sale and suits were brought by the farmers. The expensive program accomplished nothing except ill will and lack of confidence in the Federal government by those who were the victims.

In 1954 a program was instituted to eliminate the Japanese beetle along its advance into Illinois. Dieldrin was the chemical of choice in this instance. It was effective in killing the existing beetle population; it was also effective in killing much of the bird population. The poisoned beetle grubs crawled out to the surface, readily available to the insect eating birds. In the town of Sheldon, where a record was kept, virtually all of such birds were eliminated—brown thrashers, starlings, meadowlarks, grackles, pheasants, and even robins. Squirrels were also wiped out and ninety percent of the farm cats were victims as well. Nevertheless, the program continued for another eight years.

In the Indian River area of Florida dieldrin was sprayed onto the waters in an attempt to eliminate the larvae of the sandfly. After the spraying a survey by the State Board of Health reported that the fish kill was "substantially complete". An estimated 20 to 30 tons of fish, more than a million fish, were killed. The entire crab population was also destroyed.

And the list goes on with detailed documentation by Carson of the injurious effects to other fauna and flora of widespread pesticide eradication programs. But always in elegant, expressive and understandable English.

Rachel Carson had taken on the giant chemical corporations. She had accused them, among other things, of foisting on the general public toxic compounds without their consent and without adequate testing of the compounds for the effects on other fauna and flora in the environment other than the pests and weeds that they were intended to eradicate. She was soon to hear their reply.

One company tried to force Houghton Mifflin into suppressing publication of the book. They claimed that the book contained errors concerning one of their products, chlordane, and threatened suit unless proper corrections were made. The publisher declined to make the changes and no suit was filed. Others threatened to withdraw advertising from newspapers and magazines that reviewed the book favorably. In the chemical engineering trade journals commentary was derisive:

> I regard it as science fiction, to be read in the same way that the TV program *Twilight Zone* is to be watched...
>
> Her ignorance or bias on some of the considerations throws doubt on her competence to judge policy...

> Federal-state spraying continues, but it has been hampered by the adverse publicity from conservationists who claim residues of the insecticide, dieldrin, kill birds and wild animals...In any large scale pest control program in this area, we are immediately confronted with the objection of a vociferous, misinformed group of nature-balancing, organic-farming, bird-loving, unreasonable citizenry that has not been convinced of the important place of agricultural chemicals in our economy...

One of the major spokesmen for the chemical companies was Robert White-Stevens, assistant director for research and development for one of the major chemical companies. He went around the country giving speeches which extolled the virtues of pesticides and condemned *Silent Spring*. He also appeared in a television debate with Rachel Carson. The gist of his message can be garnered from the following statements that he made.

> The book is littered with crass assumptions and gross misinterpretations.
>
> The major claims in Miss Rachel Carson's book, *Silent Spring*, are gross distortions of the actual facts, completely unsupported by scientific experimental evidence and general practical experience in the field.
>
> If man were to faithfully follow the teachings of Miss Carson, we would return to the Dark Ages, and the insects and diseases and vermin would once again inherit the earth.

All the adverse efforts by the chemical companies were for nought. The scientific community, in general, supported Carson and the public was now well informed on the subject.

President Kennedy asked his Science Advisory Committee to study the whole issue. Their report, which came out in 1963, endorsed Carson's scientific findings and her service to the public.

> Public literature and the experience of Panel Members indicate that, until the publication of *Silent Spring* by Rachel Carson, people were generally

unaware of the toxicity of pesticides. The government should present this information to the public in a way that will make it aware of the dangers while recognizing the value of pesticides.

It went further in recommending that government agencies including the Department of Agriculture rethink their pesticide programs, that pesticides should be certified for safety before their first use, and that an ultimate goal was the elimination of persistent toxic pesticides.

Rachel Carson had won. In the ensuing years there have been many improvements. There has been a curtailment of highway pesticide projects, a curtailment of indiscriminate spraying of pesticides in suburban and rural areas, a curtailment of ill-conceived major pest control programs such as that for the fire ant, curtailment and control of pesticide waste dumping in streams and rivers, improvement in the procedures for registering pesticides, banning of some pesticides for home and garden, and banning of uses for DDT in general.

And all this was brought about by a single book.

Ozone Layer Depletion—Mario Molina and Sherwood Rowland

Next, we go on to a truly global environmental concern. This has to do with the stratospheric ozone layer, the protective shield which reduces incoming, and potentially dangerous, ultraviolet radiation. This protective layer has presumably existed for all of Earth history for which there has been oxygen in the atmosphere. We, and other creatures, have evolved without this added radiation from the Sun. We are now threatened with an increase in ultraviolet radiation which the Earth's fauna and flora may be poorly equipped to deal with.

The ozone layer depletion problem has had all the uncertainties that we have seen in other cases of potential toxicity. Has there been a depletion in the ozone layer? If so, what are the causes for this depletion? Has there been an increase in ultraviolet radiation on Earth? What are the natural variations in ultraviolet radiation on Earth? What harmful effects on humanity and other biota might be expected from increased ultraviolet radiation?

> **OZONE FORMATION AND DESTRUCTION**
>
> Photodisintegration and Formation of Ozone
>
> $$O_2 \rightarrow O + O$$
> $$O_2 + O \rightarrow O_3$$
>
> Ultraviolet Absorbtion and Destruction of Ozone
>
> $$O_3 \rightarrow O + O_2$$

Figure 15. *Ozone formation and destruction.*

Nevertheless, in spite of these uncertainties, actions have been taken to curtail and eventually eliminate the presumed culprits, the chlorofluorocarbons, generally referred to as the CFCs. The problem—if there was a problem—has been solved. But, the uncertainties in the ozone layer depletion problem have also led to an unusual reaction—a backlash. The whole sequence of events has been an interesting mix of science; media, public attention, and politics; and, finally, corrective actions.

First, a little science. Ozone (O_3) is a colorless gas, a molecule of which contains three oxygen atoms. The intense sunlight in the high elevations of the stratosphere produces ozone naturally by breaking down a normal oxygen molecule (O_2) into two highly reactive oxygen atoms (O), through a process referred to as photodisintegration. Then, some of the oxygen atoms combine with a normal oxygen molecule to form ozone. In turn, an ozone molecule *absorbs* ultraviolet radiation and in the process is changed back into an oxygen molecule and an oxygen atom. Some of the resultant oxygen atoms recombine to form an oxygen molecule and some combine with an oxygen molecule to form more ozone. A balance exists between production of ozone and its destruction, leading to an equilibrium concentration of ozone in the stratosphere. The absorption of ultraviolet radiation by ozone is the stratospheric process that is beneficial to humankind (see Figure 15).

> **CHLORINE AS A CATALYST IN THE DESTRUCTION OF THE OZONE LAYER**
>
> $$\underline{Cl} + O_3 \rightarrow ClO + O_2$$
>
> $$ClO + O \rightarrow \underline{Cl} + O_2$$

Figure 16. Chlorine as a catalyst in the destruction of the ozone layer.

But add an alien substance that enhances the destruction of ozone and the equilibrium is gone. And, of course, this is just what happened once humans invented the useful chlorofluorocarbons, according to many scientists. They have been in use for the past sixty years as coolants in refrigerators and air conditioners and as propellants in spray cans. CFCs are nonflammable, nontoxic, and stable, meaning that they cannot combine chemically with other substances. But once they are let loose in the air, they diffuse slowly upward into the stratosphere, where they are attacked by the intense ultraviolet radiation and break up into their primary components, one of which is chlorine.

In the ozone layer, chlorine acts as a *catalyst*, enhancing a reaction without itself being changed. And the reaction these interlopers enhance is the destruction of the ozone layer. It works like this. A chlorine atom (Cl) combines with an ozone molecule (O_3), and the result is a molecule of chlorine monoxide (ClO) and a normal oxygen molecule (O_2). But there are also oxygen atoms (O) in the neighborhood, and the chlorine monoxide molecules (ClO) react with them, producing a normal oxygen molecule (O_2) and a free-ranging chlorine atom that can head off and react again with ozone, destroying it. A relatively small amount of chlorine can do a lot of damage (see Figure 16).

So, how many times can this catalytic reaction take place? After about three years, the mean residence time of chlorine in the stratosphere, the chlorine returns to the Earth as a chlorinated compound having combined with other atoms in the strato-

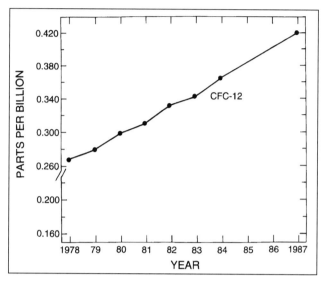

Figure 17. Measurements of atmospheric concentrations of CFC-12 over Barbados from 1978 to 1987. Concentrations are plotted in units of parts per billion. From Firor, 1990.

sphere. But during its residence time in the stratosphere one atom of chlorine can destroy a thousand to tens of thousands of ozone molecules.

And the levels of CFCs have increased in the atmosphere and stratosphere. For example, the accompanying figure shows the atmospheric levels of one of the more prevalent CFC compounds, CFC-12, as measured from 1978 to 1987 over Barbados. Over this nine year period the CFC-12 level increased from 0.26 parts per billion (ppb) to 0.42 ppb—almost doubling in less than a decade. Since then, the levels have decreased, thanks to the curtailment of the production and dissemination of the CFCs.

Now, it is sunlight that powers the creation of ozone in the delicate equilibrium mentioned. But during winter months over Antarctica, there is no sunlight. So there should be no ozone layer. And there wouldn't be except that stratospheric winds carry in ozone-rich air produced in the temperate and tropic latitudes. Yet in spite of this dynamic process, the ozone layer over Antarctica in winter has decreased by about half since 1977. The extent of the hole appears to fluctuate in size from year to year, but at times

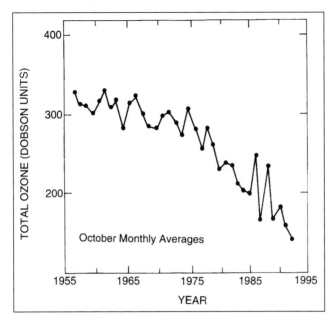

Figure 18. Historical springtime total ozone record for Halley Bay, Antarctica. From World Meteorological Organization, 1995.

it has reached an area the size of the continental United States (see Figure 18).

If there were to be a diminution in the protective ozone layer, the Antarctic at the end of the southern hemisphere winter would be the first place that it would be noticed. The stratospheric circulation over the Antarctic is more isolated than that over the Arctic, and there will be a natural diminution in ozone as the austral winter proceeds with the curtailment of sunlight. The Antarctic may be considered a precursor to what eventually may happen elsewhere. Indeed, there has, also, been a slight decrease in the ozone layer over the Arctic during the winter but none at the more temperate and tropic latitudes. Overall the global mean for the ozone layer has decreased at the rate of 1% per year from 1978 to 1987.

So, to many scientists we have a fait accompli, a cause and effect relation. The cause—an increase in CFCs in the stratosphere. The effect—the ozone hole over the Antarctic. Quod erat demonstradum, Q.E.D. But there are nagging doubts to this scenario.

First, parts per billion, the measurement unit for CFCs in the atmosphere, is an unimaginably minute measure. The CFC-12 concentrations are equivalent to one to two people out of the entire population of the world. It would be reasonable to ask if such tiny amounts of even a virulent catalyst could have a recognizable effect on the ozone layer. Well, in the stratosphere over the Antarctic the chlorine concentrations are around 1 ppb, comparable to the CFC concentrations in the lower atmosphere. The ozone concentration is around 1000 ppb, considerably greater. But with the catalytic power of a thousand or more to one there is sufficient chlorine to noticeably effect the ozone layer. Scratch one objection.

Second, are the CFCs the principal source of chlorine in the stratosphere? This is certainly not the case for the atmosphere, that body of air between the Earth's surface and the stratosphere. Evaporation of sea water puts enormous amounts of chlorine into the atmosphere in the form of sodium chloride, amounts several hundred times the CFC emissions. However, chlorine in this form is water soluble and is rapidly taken up in clouds and eventually returned to the Earth's surface as rain. Volcanic eruptions also put large amounts of chlorine into the atmosphere as hydrogen chloride, another water soluble chemical compound which is returned to the Earth as acid rain. An exception to this are the very large volcanic eruptions, such as the eruption of Tambora in 1815, which would inject large amounts of hydrogen chloride directly into the stratosphere and should have a noticeable effect on the protective ozone layer for the following one to three years. Except for the very large eruptions, scratch another objection.

Third, we are dealing with a dynamic system in the stratosphere. Without the stratospheric wind exchange between the temperate latitudes and the Antarctic, there would be no ozone layer at all during the austral winter. Isn't it reasonable to expect variations in such a dynamic system? Certainly, we see such variations in climate at the Earth's surface on whatever time scale we wish to choose from a few years, to tens of years, to hundreds and thousands of years. Isn't it possible that the global mean decrease in the ozone layer of 1% per year from 1978 to 1987, let alone the more dramatic decrease over Antarctica, may be due to natural causes? Certainly, the fact that there was an *increase* of 1% in the global mean for the ozone layer from 1961 to 1970 would argue that dynamic effects cannot be ignored.

This is just one more example of the difficulties that we often face in assessing environmental problems. We are part of an ongoing experiment, never before performed, and we lack the requisite data to predict the results. Eventually we will know if the present ozone layer depletion is caused by dynamic, i.e., natural, variations or by chemical, i.e., anthropogenic, changes or both, but only after some undefined interval of time. In the meantime, there has been a call for preventive action, which seems prudent.

The most cited effect on humans from ultraviolet radiation is skin cancer. (The radiation does not penetrate much below the surface of the skin.) Most are well aware of that danger and generally take precautionary measures to prevent such effects when exposed to intense sunlight.

The effects on lower order animals, particularly single-celled animals, and on exposed eggs and larvae may be more serious. After all, ultraviolet lamps are used to help provide a sterile environment in operating rooms and medical laboratories. Ultraviolet radiation is, indeed, lethal to most bacteria. And, the recent dieoffs of certain species of amphibians—frogs and the like—at high altitudes may have been caused by higher levels of ultraviolet radiation at these altitudes attacking the eggs and juveniles, possibly an important and dangerous warning sign. On the other hand, as of 1988, there has been no increase in ultraviolet radiation at the more temperate and tropical latitudes and this accords with the observation of no thinning of the protective ozone layer at these latitudes.

As one looks toward the future, the worst-case projection for this century is an increase of 15-20% in surface ultraviolet radiation in the Northern hemisphere. We already have a natural variation of 20% from tropical to temperate latitudes, the radiation being higher in the tropics, and, humans have accommodated themselves to that variation. Most northerners know to take adequate sunblock lotion when they go to Florida for a winter vacation. So, perhaps there is no serious problem for humans, at least.

In any event, action has been taken. Historically, the potential connection between CFCs and ozone layer depletion was first brought to the attention of the scientific community as well as the general public by Mario Molina and Sherwood Rowland, then of the University of California at Irvine, in 1974. (The two were awarded a Nobel prize in chemistry in 1995 for their pioneering effort.) The potential threat was broadcast widely, and the public reacted by deciding it could do without spray cans that use CFCs as propellants. Recognizing the futility of producing an item that people won't buy, industry was quick to follow, developing substitute propellants that they could advertise as new and improved and environmentally safe. Some states added to the effort by putting a ban on CFC-containing aerosols. But from 1974 to 1985, when there was a decrease in the use of CFCs in aerosols, the non-aerosol use of CFCs for refrigeration and air conditioning and in the semiconductor industry increased dramatically. The CFC problem was relegated to the back burner.

Then, in 1985 a British expedition announced the discovery of the decrease in the ozone layer during the winter months over Antarctica, the results of which have been discussed previously, and the game was on again. In 1987, the United States government banned the use of CFCs in spray cans and this in turn was followed by an international agreement to stabilize and ultimately eliminate the production of CFCs globally. The ozone problem—presuming, of course, that there was a problem—was on its way toward resolution.

Because of the aforementioned uncertainties in the science and in the possible health effects, there has been a backlash to the ozone layer story. Claims have been made that the whole thing was a scam that was politically motivated or that it was a hoax that was perpetrated by the atmospheric scientists to increase their research funding. Both claims are unfounded.

Nevertheless, there was a lot of media hype—overhype, if there is such a word—with consequent public alarm and reaction. A "hole in the sky" is not something to be ignored. One scientist attested that the CFCs were "attacking the Earth's immune system." And the head of the United Nations Environmental Programme declared CFC control to be the "leading ecological issue in the world." Even as late as 1992 an article in *Time* magazine started out as follows.

> The world now knows that danger is shining through the sky. The evidence is overwhelming that the earth's stratospheric ozone layer—our shield against the sun's hazardous ultraviolet rays—is being eaten away by man-made chemicals far faster than any scientist had predicted. No longer is the threat to our future; the threat is here and now. …This unprecedented assault on the planet's life-support system could have horrendous long-term effects on human health, animal life, the plants that support the food chain and just about every other strand that makes up the delicate web of nature.
> And it is too late to prevent the damage, which will worsen for years to come. The best the world can hope for is to stabilize ozone loss soon after the turn of the century.

This is colorful prose but weak on facts. One atmospheric scientist put it this way.

> On the one hand, as scientists, we are ethically bound to the scientific method, in effect promising to tell the truth, the whole truth, and nothing but—which means that we must include all the doubts, the caveats, the ifs, ands, and buts. On the other hand, we are not just scientists but human beings as well. And like most people we'd like to see the world a better place, which in this context translates into our working to reduce the risk of potentially disastrous climactic change. To do that we need to get some broad-based support, to capture the public's imagination. That, of course, entails getting loads of media coverage. So we have to offer up scary scenarios, make simplified, dramatic statements, and make little mention of any doubts we might have. This "double ethical bind" we frequently find ourselves in cannot be solved by any formula. Each of us has to decide what the right balance is between being effective and being honest. I hope that means being both.

There are dangers in trying to conduct research on a politically charged issue when there are still many scientific uncertainties.

Three lessons can be learned from all this. One, the public can act, dragging industry and government along. Two, scientists remain ignorant, nevertheless, about many matters that are of major importance, such as the physics and chemistry of the stratosphere. And, three, it sometimes doesn't take that much—in this case the production of relatively innocuous chemicals in relatively small quantities—to have a potentially global impact.

Global Warming, Part I—Roger Revelle

No environmental problem has received as much attention, as global warming. Global warming related to the emission of carbon dioxide to the atmosphere from fossil fuel burning: the greenhouse effect.

Let us start by reviewing briefly what the greenhouse effect is all about. Incoming electromagnetic radiation from the Sun is of a relatively short wavelength—what we call light. On the other hand, the heat returned from the Earth is in the form of relatively long wavelengths—what we call heat. In an atmospheric greenhouse effect, the carbon dioxide in the atmosphere is transparent to the incoming sunlight—much the same as glass in a terrestrial greenhouse. However, the outgoing heat is partially *absorbed* by the carbon dioxide and reradiated, in part, back to the Earth, thus leading to a global warming.

As indicated in the figure, the carbon dioxide in the atmosphere has the opposite effect of atmospheric dust and sulfur dioxide aerosols from large volcanic eruptions, which *reflect* the incoming sunlight, leading to a haze effect and global cooling.

Now, the greenhouse effect has, happily, been present for eons. Without it the Earth would be a huge iceball—and lifeless. Without the water vapor and carbon dioxide in the atmosphere that tend to return heat back to the Earth without preventing solar radiation from entering, the global temperature would be around -4°F, or 36 degrees below the freezing point of water and 63 degrees below the present overall global temperature, which is 59 degrees. So, we should, indeed, pause and give thanks for the global greenhouse but then look seriously at what some people expect to be an unprecedented catastrophe.

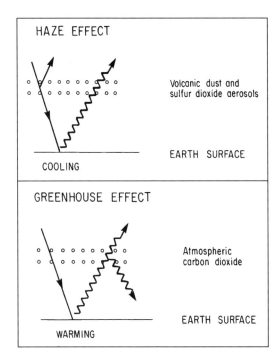

Figure 19. Haze and greenhouse effects. From Officer and Page, 1993.

First, a little geologic perspective on global temperatures. They have run a gamut far in excess of any predictions for our present day, anthropogenic effect. For example, in the Mesozoic era, 125 million years ago, the geologic time during which the dinosaurs reigned, the Earth was much warmer than it is today. The midlatitude regions were tropical in climate and the high northern and southern latitudes, temperate. Dinosaurs ranged as far north as the Arctic Sea and their remains are found in Mesozoic sedimentary strata on the North Slope of Alaska.

From this record it is not unnatural to look into the past for insight about the relationship between atmospheric carbon dioxide and global temperature. At present the atmospheric level of carbon dioxide is about 300 parts per million (ppm). One hundred and twenty million years ago carbon dioxide was six times more abundant in the atmosphere, or 1,800 ppm. And during the period that the carbon dioxide decreased to its present level, the temperature at low latitudes decreased from 82°F to 62°F. From this geologic record, one can make an estimate for the temperature increase that would result from a present day, doubling of carbon dioxide. (In the global warming discussions, a doubling of carbon dioxide is the usual reference measure taken when estimating possible global temperature changes.) We had a temperature change from 12 million years ago of 20°F with a corresponding decrease of carbon dioxide of 1,500 ppm from 1,800 ppm to 300 ppm. Thus, $(\Delta T/20) = (300/1,500)$, or $\Delta T = 4°F$ for a doubling of the present-day level of carbon dioxide. A figure that falls midway within most of the estimates from computer models.

As we look, next, into a possible anthropogenic increase in the greenhouse effect, there are, at least, three questions to be addressed. Has there been an historic increase in carbon dioxide? Has there been a global warming to date? What effects, good and bad, may be expected from a global warming?

It has been said truly many times: *what we are going through now is a large-scale geophysical experiment, the likes of which the Earth has never experienced before.* It is hard to monitor an experiment from inside the test tube. Specifically, within a couple of centuries, we will return to the atmosphere and the oceans all of the concentrated organic carbon that has been stored in sedimentary rocks for hundreds of millions of years as coal, oil, and natural gas. This sounds momentous, and it is. However, the annual emissions of carbon dioxide from these sources are actually quite small compared to the titanic amounts involved with the carbon cycle of the Earth—the natural exchanges in photosynthesis, respiration and decay, and the natural exchanges between the atmosphere and the oceans. The point is that the natural processes are operating in a kind of equilibrium, or what might be called a balanced budget. There may be no way that these processes can accommodate the additional anthropogenic emissions. And that does appear to have been the case.

The accompanying figure illustrates the essences of the global carbon cycle. About 110 million tons of carbon dioxide are taken up by photosynthetic processes and the same amount returned each year through respiration and decay. Around 5 billion tons of carbon dioxide are currently released each year as a result of the combustion of fossil fuels. Another 1 billion tons or more are released from the clearing and burning of forests, also a one way

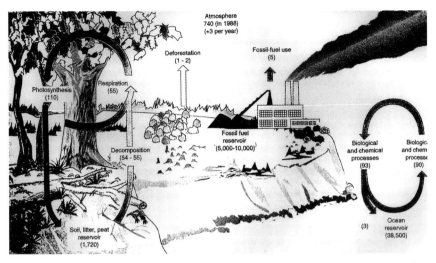

Figure 20. Annual rates of cycling of carbon dioxide among various reservoirs, expressed in units of billions of metric tons per year. From Morgan et al., 1993.

process. The ocean-atmosphere cycling, similar to that on land, is estimated to be about 90 billion tons per year. The 3 billion ton difference in the ocean-atmosphere cycling is a postulated, or cooked, figure and not a measured figure. How so? The measured atmospheric increase is only 3 billion tons per year and not the 6 billion tons, as would be inferred from the combination of the fossil fuel and deforestation figures. The only place that many scientists can think to put the remaining 3 billion tons is in the reservoir of last resort and the least known part of the carbon cycle—the oceans. It is a dilemma—but only one of several in this difficult field of climate modeling.

When we look at the first question of whether there has been an historic increase in atmospheric carbon dioxide, the answer is definitive. Yes. We know this largely due to the efforts of Roger Revelle, a former director of the Scripps Institution of Oceanography, and colleagues. Back in 1957, Revelle was concerned about the possible anthropogenic increases in atmospheric carbon dioxide from fossil fuel burning. As part of the International Geophysical Year, he instituted measurements of atmospheric carbon dioxide over Hawaii, a relatively remote, oceanic region. Those measurements continue to this day. Without

Roger Revelle. Scripps Institution of Oceanography.

Revelle's foresight we would not have the record that has brought the problem of potential global warming to such prominence.

Over the thirty-year period from 1958 to 1988, carbon dioxide in the Hawaiian atmosphere has increased from 315 ppm to 350 ppm, in all an increase of 35 ppm. The large annual variations shown in the Hawaii graph reflect the uptake of carbon dioxide from the air by plants as they grow in the spring and summer and the return of carbon dioxide to the air as they decay in the autumn and winter (see Figure 21).

At this point we run into one more dilemma in the global warming problem. It goes like this. So, we have a record of the increase in atmospheric carbon dioxide and we can reasonably attribute it to anthropogenic causes, what sort of increase may we expect in the future and by what date may we expect there to be a doubling of atmospheric carbon dioxide?

One assumption might be to presume that the present rate of increase will continue on into this century. This implies that a doubling would take 270 years from the 1958 base date, i.e., (30/35) x 315. That is, there would be a doubling by around the year 2230 A.D.

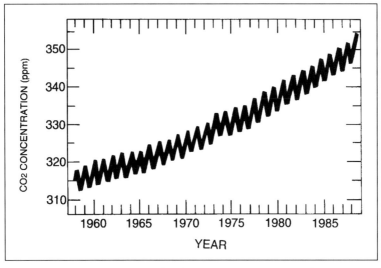

Figure 21. The concentration of carbon dioxide in the atmosphere as measured at the Mauna Loa Observatory, Hawaii, from 1958 to 1988. From Morgan et al., 1993.

Now, most of the individuals who do climate modeling assume that atmospheric carbon dioxide will double by the middle of this century, that is, within about 90 years of a 1958 base date. That is quite different from the directly extrapolated figure of 270 years. What the modelers are doing is assuming that there will be a substantial increase in the rate of fossil fuel consumption as we proceed into this century—an exponential increase corresponding to the present exponential increase in global population and energy consumption.

The modelers are taking a worst-case scenario of global warming and it is quite proper to do just that. But that fact gets lost in the background noise of media reporting on global warming. If we are able to curtail our consumption of fossil fuels to, say, the 1958 level, the 270-year doubling figure would be applicable and many of the dire consequences predicted for the middle of this century would not occur. And there are very good reasons for doing just that because even at our present rate of fossil fuel consumption, we will run out of oil within 100 years.

Next, on to a brief look at what we already instinctively know.

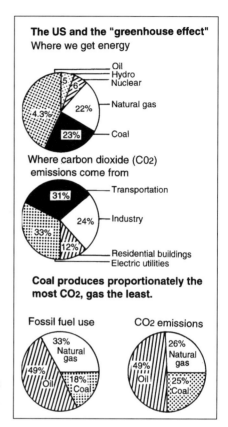

Figure 22. Sources of carbon dioxide in the atmosphere. From Dumanoski, 1988.

What are the sources of all this anthropogenic carbon dioxide? From the upper pie diagram we see that we get 89% of our energy from fossil fuel burning of oil, coal, and natural gas and in that order of importance. In the middle diagram we see that carbon dioxide emissions come first from electric utilities which are largely coal burning; then, in importance, from transportation which is largely oil burning; and, third, from industry. The lower diagram shows the percentage contribution from each of the fossil fuels.

Now, on to the second question: has there been an observable increase in global temperature related to anthropogenic causes? The answer is perhaps. In the Northern Hemisphere, the average temperature rose about one degree Fahrenheit from 1900 to 1940. Then, it decreased by half a degree between 1940 and 1970—with

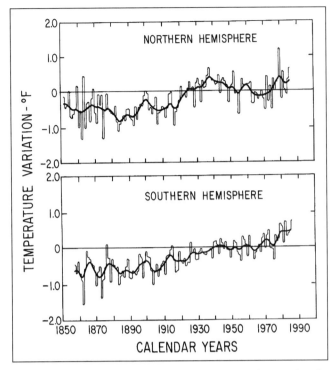

Figure 23. Historical records of temperature change for the Northern Hemisphere and Southern Hemisphere. From Michaels, 1990.

predictions that we were headed into the next ice age. Since then, it has risen half a degree, returning the Northern Hemisphere to its 1940 state. In the Southern Hemisphere, the pattern is somewhat different. Rather than the ups and downs, the temperature has risen quite steadily by a total of one degree Fahrenheit over the past one hundred years. The warming trend in recent years is certainly there in both hemispheres but whether it has been caused by anthropogenic or by natural causes remains debatable. After all, we did have the Little Ice Age of 300 years ago preceded by the Medieval Warm Period in the 1200s, with temperatures down in the case of the Little Ice Age and up in the case of Medieval Warm Period (see Figure 23).

As we go on to look at projected temperatures for tomorrow, it should be appreciated—at the outset—that it is difficult to make accurate predictions. Climate modeling is filled with

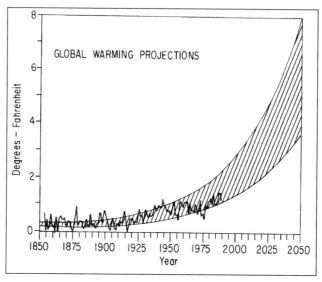

Figure 24. The range of global warming predicted by various computer models (diagonal region) superimposed on the historical record of temperature change. All the models predict that warming will accelerate significantly in coming decades. From Jones and Wigley, 1990.

vagaries. For example, what will solar radiation changes be over the next hundred years? They will certainly affect any warming trend. How will cloud cover change with increased warming and what will its water content be? In one model the inclusion of cloud cover lowered the anticipated global warming from 10°F to 4°F. And, what are the overall effects of that great heat sink, the oceans? In the past models had a range of 2-11°F for a doubling of atmospheric carbon dioxide; as of 1999, a consensus toward a more narrow range of 4-6°F was emerging.

The accompanying figure shows a summary of some model calculations as of 1990. As stated in the legend to the figure, the modelers project that the warming will accelerate significantly in coming decades. In other words, that the doubling time for carbon dioxide will be considerably shorter than that given by a direct extrapolation of the present carbon dioxide level—that, indeed, the combustion of fossil fuels will increase substantially in the years ahead.

In these models atmospheric carbon dioxide has doubled by the year 2050 with a projected global increase, on the low side of 4°F to an amount, on the high side, of 8°F. Note, on the one hand, the projected increase in the rate of temperature rise, in accordance with the model assumptions. Note, on the other hand, that the historic temperature trend from 1910 to 1970, if attributed to a greenhouse effect, is at the lower end of the projections.

There are two other projections that come out of the models. First, there is general agreement that the temperature will rise more near the poles and not very much at all near the equator. The polar rise will be about three times the global average.

Second, the journey from one equilibrium temperature state to another at a higher temperature will not be a smooth trip. There will be lots of hills and valleys. Projections suggest that the weather will be both *hotter and colder* and *wetter and drier* than in previous years. Many suggest that we are already seeing such effects—the summer drought conditions of 1988 throughout the Midwest, the heavy snowfall of the winter of 1995-1996 in the Northeast, and the heavy rain and snow of the winter of 1996-1997 in the Northwest. And these large fluctuations could be more devastating than the global temperature rise itself.

We have reached the third question: what effects may be expected from a global warming? Let us follow the worst case scenario of the climate modelers.

Many in northern climes would welcome a little warmer weather. More northerly lands would become more arable. Canadian farmers would get richer. A place like Iceland would have the climate of present-day Scotland and would be able to grow as much hay and barley as Scotland. As has been discussed previously, carbon dioxide is a necessary part of the photosynthetic process. For certain crop plants like barley, wheat, rice, and potatoes a doubling of carbon dioxide would lead to an increase in crop yield by 30% or more. Some forecast a net gain for agriculture, though many adjustments will be necessary.

On the other hand, places like the American Midwest, the bread basket of the United States, would become more arid and less productive, perhaps calling for the growth of different crop plants.

In general, warm-weather plants will move poleward as will the animals associated with them—if they can. There is great doubt that many trees, for example, can migrate fast enough. At the end of the last Ice Age, beech trees were able to follow the ice northward at a rate of about 12 miles a century. But if carbon dioxide doubles within this century, beech trees would have to move 300 miles; and no one thinks they can do that. Most adult trees can withstand a considerable rise in temperature, but saplings cannot. Many forests will simply vanish after the old trees have died. Parks and refuges designed to preserve particular ecosystems will see them unravel, with some species able to move northward, others not. Presumably, rates of extinction will increase above what would take place without the added stress of global warming.

The global increase will, as noted, be greater in the high latitudes of the Arctic and Antarctic. A 6°F rise globally would be felt in the high latitudes as 18°F. The effects on the world's oceans could be profound. Melting of the Antarctic ice sheets and the simple expansion of seawater as it warms up could provide a rise in sea level of three feet. Marshes would be inundated, not having enough time to move inland as the sea rose. Virtually all such places would vanish, as would most of what remains of the world's beaches. Coral reefs would probably not be able to build themselves fast enough to maintain the proper distance below the ocean surface. Many barrier islands would disappear under the sea. Coastal cities like Miami will have to build expensive dikes, as in the Netherlands, or become much smaller.

Not a very pleasant scenario.

First off, there are a number of vagaries in the predictions. One is the assumption of an accelerated increase in atmospheric carbon dioxide over this century—a substantial increase over what has been observed to date. Then, there is the problem that half the anthropogenic carbon dioxide emitted does not end up in the atmosphere and remains unaccounted for in a carbon cycle budget—the oceans being taken as a convenient but, as yet, unproved sink. There are, also, the unknowns in climate modeling itself; we are far from having an adequate and sufficient understanding of how the global climate system works.

With this background some scientists have argued that we should continue with research, modeling, and data collection—a procedure that most people agree on—but that this should be done *before* taking precipitous action at great expense for a problem that may not exist.

Other scientists urge us to adopt what might be called a *precautionary principle*. The concept is that when dealing with risky, far-reaching, and often ultimately irreversible environmental problems, the safest course is to take informed action, even before there is enough scientific knowledge to justify that action completely. That was done with the ozone layer depletion problem. But there is a big difference between the two problems. The ozone layer problem required the elimination of a relatively unimportant compound, the CFCs, for which substitutes could be easily developed. The global warming problem requires a reevaluation of our whole concept of energy consumption.

First and foremost in curtailing the global warming problem is fossil fuel efficiency. This means curtailing anthropogenic carbon dioxide emissions by more efficient transportation systems, removal of carbon dioxide from smokestacks and vehicle discharges, and changing from inefficient coal burning to natural gas. These are relatively easy changes to envisage with improved technology. In the long run we need to switch to alternative fuels, another change that is technology dependent. Neither solar nor nuclear energy discharge carbon dioxide to the atmosphere.

There is also a stronger reason to go along with this precautionary principle. Global warming is directly related to and caused by the two most important global environmental problems that we face—population growth and exhaustion of nonrenewable natural resources. These three are interlinked, and they must be addressed through patient and concerted efforts by all of us.

These latter two, population growth and alternative energy sources, are the subjects of the last two chapters in this book.

Global Warming, Part II—Wallace Broecker

There is another aspect of the global warming experiment that deserves separate attention. The biggest worry may come not from what we understand about the global climate system and

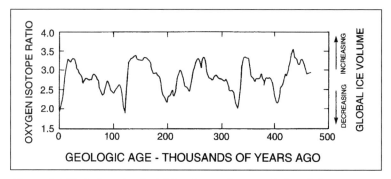

Figure 25. Variations in glacial ice sheet volume as inferred from oxygen-isotope measurements versus geologic time. From Imbrie and Imbrie, 1979.

can model numerically but rather from what we do not understand and are unable to model. As succinctly stated by Wallace Broecker, an Earth Scientist at Columbia University, *the biggest worry may be from the unpredictable—what we may inadvertently trigger.*

To get a better understanding of this danger, it is useful to go back and examine climactic events of the recent and geologic past. They are quite varied and some are quite dramatic. We have made considerable progress in our scientific understanding of the causes of these events but are, still, far from what we would like to know and understand.

Geologically speaking, we are living in an Ice Age. At present we have an interglacial period of temperate climates. But only 20,000 years ago we were at a peak of a glacial cycle with a mile or more thickness of ice covering Canada and a somewhat lesser amount covering Scandinavia. All this water came from the oceans with a consequent drop in sea level of 300 feet, exposing all of the world's continental shelves.

There have been several glacial cycles over the past million years or so. The accompanying figure shows the variation in global ice volume over the past 500,000 years. The glacial-interglacial cycle is around 100,000 years and the glacial volume in each cycle has built up gradually over time with a rapid decrease to the next succeeding interglacial (see Figure 25).

(This record comes to us through the magic of geochemistry.

Briefly, the reasoning goes as follows. There are two stable isotopes of oxygen, one with an atomic weight of 18 and the other with an atomic weight of 16. When water evaporates from the ocean, the heavier isotope is preferentially left behind. As this water goes into building up ice sheets, the oceans become relatively enriched in oxygen-18. In turn oceanic oxygen is taken up in building the calcium carbonate shells of plankton. The expired plankton sink to the bottom and become a record of the variations of the ratio of oxygen-18 to oxygen-16 through geologic time.)

The waxing and waning of the continental ice sheets is attributed to an external forcing function having to do with variations in solar radiation on the Earth. The Earth revolves around the Sun in an elliptical orbit known as the plane of the ecliptic. The Earth also rotates around its own axis, which defines the equatorial plane. The plane of the ecliptic and the Earth's equatorial plane are set off from each other at an angle called the tilt. Both the shape of the elliptical orbit of the Earth around the Sun and the tilt between the ecliptic and equatorial planes vary slightly over time, given the perturbing gravitational effects of the other planets in the Solar System. The orbital variations have a period of 100,000 years; those of the tilt have a period of 40,000 years. In addition, the rotational axis of the Earth itself precesses, or wobbles, with a period of 22,000 years. The wobble comes about because the Earth is not a perfect sphere but instead an oblate ellipsoid, being wider (at the equator) than it is high (from pole to pole), and because of the Moon's gravitational attraction on this ellipsoid shape. These three astronomical variations are called the Milankovitch cycles, named after the Yugoslavian mathematician Milutin Milankovitch, who made the necessary, tedious calculations in the early 1900s before the computer era.

Unfortunately, the Milankovitch cycle theory cannot be the complete explanation. Specifically, the 100,000-year period is the weakest, by far, of the three astronomical cycles whereas this period is the strongest in the oxygen-isotope record. Also, the Milankovitch cycles cannot explain the sawtooth nature of the ice record with a gradual build up and a rapid decline. We must look elsewhere for a resolution and this remains one of the major unknowns of global climatology.

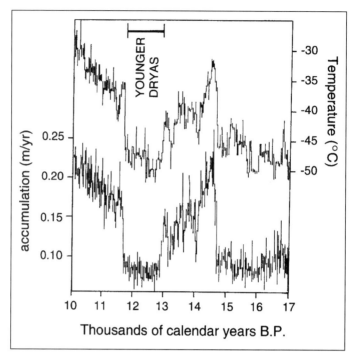

Figure 26. Comparison of temperature and precipitation accumulation for the Younger Dryas event from the Greenland ice cores. From Jouzel et al., 1997.

Ice ages are not a common occurrence in the geologic record. A reasonable question is why do we have one now? The answer lies in the levels of the greenhouse gases. There are two stable equilibria for the Earth—all ice or no ice. As we have mentioned previously, without any greenhouse gases the Earth would be a huge iceball. Throughout most of geologic time there has been several times the present level of carbon dioxide in the atmosphere. The high latitudes have had a temperate climate and have been ice free. The only geologic time with levels of carbon dioxide as low as ours was 300 million years ago, and it was the only other geologic period with significant glaciation.

Looking with a finer-tooth comb at our present Ice Age, we see climactic events within each glacial-interglacial cycle. They are referred to as stadials for cold periods and interstadials for warm periods. Some twenty-four such stadial-interstadials have been identified during our most recent glacial epoch alone. We are not at all sure of the origin of these events.

We are interested here, however, in one of the *more recent* and, certainly, the *most dramatic* of stadials. It occurred during our present interglacial period, only a few millennia ago; and unlike the previous stadials, we think we know the cause of this climactic event. It is referred to as the *Younger Dryas*. The dryas is an Arctic flower prevalent in Europe during this cold period. The term "Younger Dryas" is used to distinguish this stadial from two preceding but lesser stadials, the "Older Dryas" and the "Oldest Dryas".

Ice cores from Greenland give a sharp and distinct documentation of the Younger Dryas. From the accumulated precipitation these cores give a year-by-year record of the variations in atmospheric temperature at the site. The onset of the Younger Dryas began 12,900 calendar years ago. In a matter of less than fifty years the temperature dropped by 7-10°C (13-18°F), back to glacial levels. The Younger Dryas cold period lasted until 11,700 calendar years ago at which time the temperature returned to its previous level in the same short time interval of less than fifty years (see Figure 26).

(Once again the magic of geochemistry comes into play in providing this documentation. Converse to the evaporation process, condensation will favor the heavier, oxygen-18 isotope. As water vapor is transported from the tropics to the high latitudes, successive cooling and precipitation will lead to depletion of oxygen-18 in the remaining water vapor. The greater the fall in temperature, the more condensation will have occurred, and the lower will be the heavy isotope concentration relative to the original water vapor. Isotopic concentration can, thus, be considered as a primary function of the temperature at which the condensation took place. In addition, the annual variation in the ratio of oxygen-18 to oxygen-16 from cold, winter conditions to warm, summer conditions provides the means for dating the cores year-by-year, much the same as is done with tree rings.)

As might be expected, there were enormous floral and faunal changes in the Northern Hemisphere associated with this sudden change in climate. As with the previous glacial times, there was less rainfall and an expansion of arid conditions throughout the

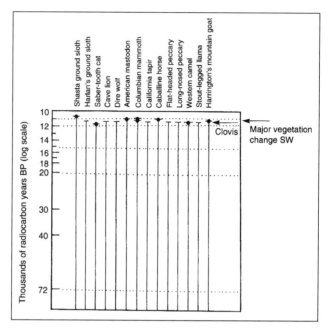

Figure 27. Chart showing estimates of latest survival of selected extinct mammalian species in North America. From Stuart, 1991.

region. In North America and Europe grasslands gave way to a sparse vegetation cover. The onset of the Younger Dryas and its severe cold conditions was also the time of migration of paleolithic peoples to North America—the migration of Clovis Man from the Bering Sea region to the American southwest.

The most striking event associated with the Younger Dryas, however, is the disappearance, i.e., extinction, of much of the late Pleistocene mammalia. It is the most recognizable extinction event in geologic history since the demise of the dinosaurs some 65 million years previously. The list of North American bestiary that became extinct is vast including mastodon, mammoth, saber-tooth cat, sloth, lion, wolf, tapir, horse, peccary, camel, llama, and goat. As shown in the diagram, the close synchronicity of the last occurrences of all these animals with the onset of the Younger Dryas is most remarkable. The same agreement holds for the disappearance of the Woolly mammoth and Irish elk in Eurasia (see Figure 27).

Two comments on the diagram. One, the uppermost or "highest occurrence" of a given species in the stratigraphic record may not always be associated with its "last appearance" on Earth. If there are few sampling sites available with identifiable bones for a given species, its last appearance may be somewhat later than its highest occurrence in the geologic record. Two, most of the dating for these and other specimens of this general geologic age is by the radiocarbon method. In that, the amount of radioactive carbon, carbon-14, still present in a sample provides an estimate of its age. It is a relative age determination method which has to be calibrated against an absolute method. When this is done, it is found that the onset of the Younger Dryas has a radiocarbon age of 11,100 years B.P. (before present) and a calendar age of 12,900 years B.P.

Those who have studied these extinctions have attributed them to either the vegetational changes of the Younger Dryas or to overkill from the new invader, paleolithic man, or to both. We subscribe to the combined environmental change and overkill hypothesis as espoused by Anthony Stuart of the Castle Museum in Norwich, England, viz.,

> Combined overkill and environmental change hypothesis. Although neither factor alone, is satisfactory, an hypothesis involving human predation on a population already stressed by climate/environmental changes can account for most of the pattern of extinctions, as currently known.

The principal point is that both the climate/environmental change and the appearance of paleolithic man are Younger Dryas related events.

For our present world with its overpopulation and limited food resources the reappearance of a Younger Dryas event would be catastrophic. And remember, the Younger Dryas came swooping in in less than fifty years; there would be little time to adapt.

We think that we understand the origin of the Younger Dryas, thanks to the efforts of Broecker. It is related to changes in *oceanic* circulation rather than to changes in *atmospheric* circulation.

His reasoning goes like this. There is a gigantic oceanic system

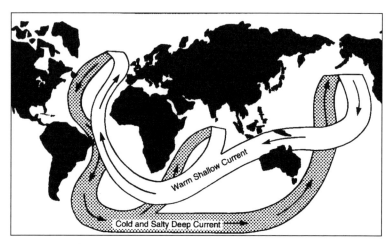

Figure 28. The great ocean conveyor carries warm water to the North Atlantic as a surface flow and cold water to the South Atlkantic as a deep flow. From Broecker, 1997.

that transports heat from the Southern Hemisphere to the Northern Hemisphere. In the oceanic region between Greenland, Iceland, and Norway the surface waters are both colder and more saline than adjacent waters to the south. They are colder because of the northern latitude and more saline because evaporation exceeds precipitation in this region. These denser waters sink to the bottom and flow south to the Antarctic as an identifiable water mass known as the North Atlantic Deep Water. There these waters join the Antarctic Bottom Water, eventually being transported to the north in the Indian and Pacific Oceans. The return flow is of less dense and *warmer* surface waters from the equatorial Indian and Pacific Oceans to the South Atlantic Ocean and ultimately to the North Atlantic Ocean where with the help of the Gulf Stream this heat is distributed to North America and Europe (see Figure 28).

Going back, now, to the period in which the ice sheet that covered Canada and the northern portion of the United States was melting, the meltwater flow was initially down the Mississippi River. At the time of the onset of the Younger Dryas the ice sheet had melted back sufficiently that the flow was diverted to the St. Lawrence River. It is conjectured that this addition of freshwater to the formation region of the North Atlantic Deep Water was sufficient to lower the water density such that no Deep Water was formed. Thus, the oceanic "conveyor belt" shut down and the Northern Hemisphere plunged into the Younger Dryas cold period.

This is just one example—one that we think we understand—of the complex relations that can exist in our coupled oceanic and atmospheric climate system. Going forward in time, there are many other examples of climactic events that have been smaller than the Younger Dryas but which, nevertheless, have affected humanity. For example, around 8,000 years ago we had a warm period, known as the Holocene Climactic Optimum. Regions such as Mesopotamia were fertile. One of our earliest civilizations developed there, in a region which is now largely arid. From about 1450 to 1850 A.D. we had what is called the Little Ice Age. Temperatures in Europe and North America were lower than today. The Thames frequently froze over and there were shops and stalls on the ice across the river. The Little Ice Age was preceded by the Medieval Warm Period. The Vikings established settlements on both the east and west coasts of Greenland, settlements that had to be abandoned when the Little Ice Age came along. We are not at all clear as to the origin of these relatively recent climactic events.

Finally, let us now look at the present with this knowledge of the coupling of global oceanic and atmospheric systems and the hemispheric transport of heat via the oceanic conveyor belt. The Younger Dryas occurred during a period of global warming from glacial to interglacial times. We are now, also, in a period of global warming related, in this case, to anthropogenic carbon dioxide emissions. Are there similarities between the two periods in what may be inadvertently and unexpectedly triggered? The answer is maybe.

In 1975 there was a joint Canadian and American Arctic expedition, known as AIDJEX, in which camps were manned on the Beaufort Sea ice pack to permit various meteorologic and oceanographic measurements to be taken over a period of time. The Beaufort Sea is the portion of the Arctic Ocean north of Alaska. At the end of the melt season in October an amount of 0.8 meters of freshwater had accumulated beneath the ice pack. The measurements were repeated in 1997 in a program known as

SHEBA. At the end of the melt season in October of 1997 an amount of 2.0 meters of freshwater had accumulated beneath the ice pack. Miles McPhee of McPhee Research, Timothy Stanton of the Naval Post Graduate School, James Morison of the University of Washington, and Douglas Martinson of Columbia University note, with much caution, in their scientific article on this latter program that "a 2.5-m [meter] melt in one season would *eliminate* most of the ice pack". (The italics have been added.) We may be dangerously close to the mythical Open Polar Sea that explorers Elisha Kent Kane, Charles Francis Hall, and George Washington DeLong sought in vain during the 1800s.

The increased melting in the Arctic is not attributed directly to a greenhouse warming but rather to an atmospheric change which has brought warmer air into the region. It is referred to as the Arctic Oscillation and is a naturally occurring phenomenon with a period of years to decades; however, since 1980 it has remained in a strong positive phase with warming conditions in the Arctic. One, unfortunately, ends up with the usual question as to whether this most recent development in Arctic atmospheric circulation is due to natural causes (a cyclical variation of some sort) or to an anthropogenic trend (a one-way street related to increased carbon dioxide in the atmosphere). Certainly, time will tell—when it may be too late; and a better understanding of the causes of the Arctic Oscillation would be of immense benefit.

A few simple calculations. An additional 1.2 meters of freshwater was produced in 1997 than in 1975. With a summer ice area for the Arctic of 5.2 million square kilometers, this gives an additional yearly freshwater production of $R = (5.2 \times 10^6 \times 1.2 \times 10^{-3})/22 = 0.3 \times 10^3$ cubic kilometers per year. The only deep water passage out of the Arctic Ocean is the Fram Strait between Greenland and Iceland. This Arctic water exits into the North Atlantic Deep Water formation region. For comparison, the average yearly production of freshwater during the glacial melting can be calculated from the observed rise in sea level, 110 meters; the area of the oceans, 361 million square kilometers; and the total observed time of melting of 12,000 years. $R = (361 \times 10^6 \times 0.11)/(12 \times 10^3) = 3.3 \times 10^3$ cubic kilometers per year. The present conditions are an observable fraction of the glacial conditions.

Big question. Are we already in a state of decreased production of North Atlantic Deep Water and slowing down of the oceanic conveyor belt? The two items discussed below would argue in the affirmative. Albeit all quite speculative.

Coral bleaching, i.e., the loss of the microscopic algae which inhabit the coral cells and give the coral its color and provide it with nourishment, and subsequent coral mortality have been of concern in the tropical waters of the world since, at least, 1983. In 1998, coral reefs around the world appear to have suffered the most extensive and severe bleaching and mortality in modern record. The equatorial Indian Ocean has been the most severely impacted with 70-80% of coral morality. It has been an ecological disaster.

The mortality has been attributed to a rise in maximum summer temperatures of 1-2°C (2-4°F). That temperature increase is much greater than any predictions from the direct effects of global warming. However, if there has been a slowing down of the oceanic conveyor belt, there would be less heat transported away from the equatorial Indian Ocean and a consequent rise in sea surface temperature there.

The Sahel is the region of Africa that extends from Mauritania in the west to Somalia in the east. The word "Sahel" derives from an Arabic word for "shore". And it is a kind of shore, or transition zone, between the dry terrain of the Sahara to the north and the well-watered grasslands of central Africa to the south. The Sahel does have a rainy season and the amount of rainfall can vary substantially from year to year leading to either fertile or drought conditions. However, over the past thirty years since the late 1960s there have been persistent drought conditions which have increased in severity year after year. As a consequence, starvation has been rampant with human deaths measured in the hundreds of thousands to millions.

Over this same time period there has developed an increasing anomaly in sea surface temperature between the equatorial South Atlantic and equatorial North Atlantic of up to 0.8°C (1.4°F). When C.K. Folland, T.N. Palmer, and D.E. Parker of the Meteorological Office in England include this anomaly in their global atmospheric circulation model, they find that increasing sea surface temperature anomalies to the levels observed modulate the summer rainfall in the Sahel through changes in equatorial circulation. The greater the sea surface temperature anomaly, the less the rainfall. Again, if the oceanic conveyor belt has slowed down, there will be less heat transport from the equatorial South Atlantic to the equatorial North Atlantic and a resultant increase in the temperature difference between the two.

We make the following observations from this discussion. (1) There have been a vast number of climactic events in our recent past. (2) We have an understanding of the causes of a few of these events but by no means all of them. (3) A major climactic event—the Younger Dryas—occurred only twelve millennia ago. From the Greenland ice cores there was a decrease in atmospheric temperature of 13-18°F and the decrease occurred rapidly over a period of less than fifty years. The Younger Dryas cold period lasted for a thousand years and, then, the return to a more equitable climate was as rapid as the descent to the cold conditions. The Younger Dryas is associated with our latest extinction event, that of the Woolly mammoth, American mastodon, Saber-tooth cat, and cohorts. The cause of the Younger Dryas is seen to be in a cessation of the oceanic, conveyor belt system which transports heat from the South Atlantic to the North Atlantic Ocean. (4) In the recent past and continuing to the present there has been extensive coral mortality in the tropical Indian Ocean, the persistence of arid conditions in the Sahel, and disturbing changes in the ice pack covering the Arctic Ocean.

We have given a scenario here of the possible interrelations among these three recent observables; others could give other scenarios. The important point is that the Arctic, Sahel, and Indian Ocean changes are unusual events which are very difficult—if not impossible—to explain in terms of natural causes. We are fooling around with a global climactic system that we little understand. In our great geophysical experiment of adding substantial amounts of carbon dioxide to the atmosphere, we may trigger dire consequences that we have been unable to predict. *The only prudent course of action is to, indeed, follow that of the precautionary principle and change our energy consumption practices from the fossil fuels to alternative and nonpolluting sources.*

Green Revolution—Norman Borlaug

In the previous two sections we addressed the problem of global warming—the data, the vaguaries in our scientific understanding, and the potential effects. An obvious, but nevertheless important, corollary is that global warming is related to our profligate use of the fossil fuels. The more fundamental problem is the need to develop through science and technology alternative energy sources. And that is the subject of Chapter 9.

A similar comparison can be made to the subject of this section—global food production. As with global warming, it is not the fundamental problem. It leads directly to the problem of the global population explosion. And that is the subject of Chapter 8.

So, on to global food production and how the population problem drives the food problem. A fitting place to start is the so-called "dismal theorem" of Thomas Robert Malthus. It is the center piece and raison d'être for Chapter 8 but let us preview it here.

The gist of Malthus's theorem is that food production will increase *linearly*—acre by acre. And further that there is a finite limit as to how much food can be produced—limited by the improvements in agricultural production procedures and the amount of land on which crops can be grown. On the other hand, population can, and indeed has in the recent past, increase *exponentially*. For example, a population doubling every twenty-five years, a human reproduction cycle, will lead to a population increase from one million to one billion in two and a half centuries.

Malthus's thesis is that at some point in human history population growth will outstrip food production. Unfortunately we have reached that stage in human development. Malnutrition and starvation are rampant in many of the lesser developed countries. Albeit, at the same time with overproduction in some of the more developed countries, such as the United States.

When we look at global agriculture, there is both some good news and some bad news. The good news is that between 1950 and 1990 world grain production tripled, helping to reduce hunger and starvation around the world. It is referred to as the *Green Revolution*. The Green Revolution began in improved agricultural practices in the more developed countries and was then carried over to the lesser developed countries (see Figure 29).

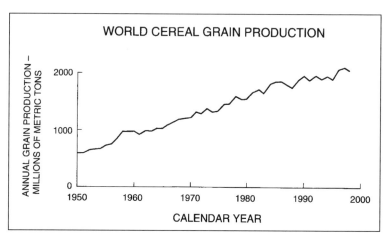

Figure 29. Total worldwide production of cereal grains from 1950 to 1998. Data from United Nations Food and Agriculture Organization.

The Green Revolution came about through a number of agricultural production improvements. One of the more important was the development of hybrid crops, i.e., genetically engineered new types of wheat, corn, and rice. In this, the male part of one grain that has some desirable features is germinated with the female part of another grain that has other desirable features. The net result is to produce a new grain that is faster growing, hardier, more resistant to indigenous pests, more suitable to higher altitudes, and/or more suitable to regions with less rainfall. Such hybrids require meticulous gathering of the grain spores but can advance from testing to field production is a matter of a year or so.

Along with hybrid crops are the equally important improvements in extended mechanization to allow vast regions to be cultivated with limited human labor, such as the American Midwest; the use of artificial fertilizers to replace the nutrients in the soil that are removed whenever crops are harvested; the use of pesticides; and the use of irrigation.

Exporting the Green Revolution from the United States to other countries around the world involved many people. One of the more notable among them was Norman Borlaug, a plant pathologist and geneticist. His first efforts were in Mexico, specifically, in wheat production. He developed hybrids suitable to

Norman Borlaug. Atlantic Monthly.

growing conditions in Mexico and showed that it was possible to speed up crop production by harvesting two generations of this new wheat every year—one in Sonora, close to sea level, and the other in the mountains near Mexico City. Wheat production improved so dramatically that by 1948, instead of importing half its wheat to feed its people as before, Mexico had become self sufficient. All of this was accomplished within a mere four years. He and his associates then went on to develop similar grain hybrids for the benefit of countries in South America, the Middle East, Asia, and Africa.

In 1970 Borlaug was awarded the Nobel Peace Prize for his part in the Green Revolution, which allowed a greater measure of peace and prosperity throughout the lesser developed countries. In his Nobel acceptance speech, he made it very clear that an adequate food supply, although essential to a stable world order, was only a first step and that there was much left to be done. He identified population growth as the biggest problem facing humanity. In essence, the Green Revolution gave the world a breathing peri-

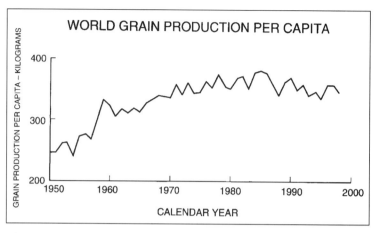

Figure 30. Per capita grain production of cereal grains from 1950 to 1998. Data from United Nations Food and Agriculture Organization.

od in which to solve its population problem; and the world has not done so.

The bad news is shown in the graph of per capita, rather than total, grain production from 1980 onward. The per capita grain production increased by about 40% from 1950 to 1980, a notable achievement of the Green Revolution. Unfortunately, since then, the per capita grain production has leveled out and begun to decrease. The hare of population growth with its exponential increase has now risen to overtake the tortoise of agricultural production with its linear increase. In a sense the Green Revolution can be considered to have provided temporary relief to the global food problem, although it should be noted that the per capita grain production is, nevertheless, higher today than it was in 1950 see (see Figure 30).

Can this trend be reversed? The answer is that it cannot be done through increases in agricultural production alone. It will require, rather, a concerted effort at population stabilization. However, let us look at some of the agricultural aspects here.

First and foremost, we are near the limit of increasing agricultural production by cultivating new land. Most of the arable land is already under production. What is left is marginal land—at high altitudes or in drought susceptible regions—that can produce little grain. We have overrun the world—a sobering thought.

On the other hand, we can reasonably expect that the Green Revolution will continue in the lesser developed countries through the introduction of hybrids appropriate to the climate and population and through the spread of knowledge in mechanization, irrigation, and the use of fertilizers and pesticides.

Then, there is the simple matter of transportation. Many crops—such as those in the Aral Sea region—are not harvested or go to waste after harvesting because the transportation system to deliver them where there is a need does not exist. Certainly, improvement in this regard can be made to alleviate global hunger.

Also, there is the use that is made of the grains produced. In this country about 90% of all grain is fed to livestock that provide meat, milk, and eggs to American consumers. This is an inefficient use of grain through the introduction of a secondary processing procedure before ultimate consumption by humans. The ratio of energy input to protein output is 2.8 for corn and 2.3 for wheat as compared with 25.0 for beef. A change in American dietary practices would free up vast amounts of corn and wheat for direct consumption elsewhere. And, of course, this is happening, not for such grandiose notions as relieving global hunger but rather for more immediate considerations of heart disease and cholesterol levels.

Finally, a few comments on international food relief—Somalia, Rwanda, and others. Most people view it as a proper humanitarian effort to prevent people from dying. That is admirable but it should also be viewed as temporary relief, and only supplied when there has been a breakdown in a region's food supply because of natural or anthropogenic disasters. The important, long-term consideration is to help countries grow their own food and control their own population growth. Without that, food relief—if a permanent prop—will only lead to further population growth and to further premature deaths from starvation. Hard words, but unfortunately true.

As one goes on to an examination of world fishing, much the same picture emerges. On the up side, the total fish catch has increased substantially over the period from 1955 to 1997. The down side is that the estimates for total sustainable yield from all

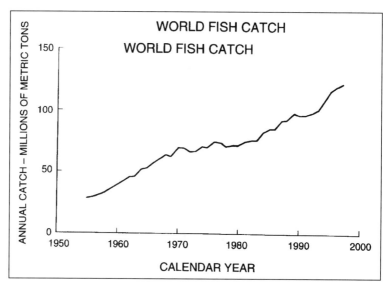

Figure 31. World fish catch from 1955 to 1997. Data from United Nations Food and Agriculture Organization.

the world's fishing grounds is 100 to 110 million tons. We have already exceeded that limit (see Figure 31).

According to the United Nations Food and Agriculture Organization, thirteen of the world's seventeen major fishing zones had either reached or exceeded their estimated maximum sustainable yield for commercially valuable fish by 1993. In nine of these areas yields have been dropping sharply for several years.

In some cases the situation is even more dire. Specifically, once prolific fishing grounds no longer exist. This is the case for herring off Norway, cod and haddock on Georges Bank, and anchovy off Peru.

All three collapses are related to a combination of overfishing and natural variations. By overfishing we mean, simply, the taking of so many fish that there is insufficient breeding stock left to maintain the population. The associated natural variations are either wide changes from year to year in recruitment to a fish population as in the case of herring, cod, and haddock or wide changes in food for the fish as in the case for anchovy.

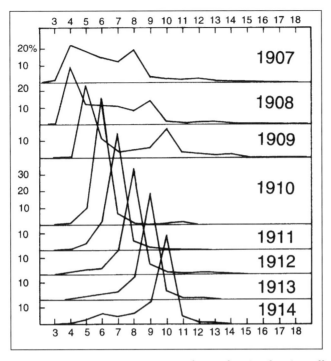

Figure 32. Composition in point of age of spring herring off Norway from 1907 to 1914. From Hjort, 1914.

Let us start with herring. The demographic chart shows the importance of natural variations and, in this case, the amazing difference in age for herring caught off Norway for the period from 1907 to 1914. (Fish ages can be determined from their scales, much the same as with trees.) What one sees is that the year-class of 1904 and secondarily the year-class of 1899 dominate the entire catch. In 1908 these fish were four years old; in 1914 they were ten years old. Without the 1904 year-class there would have been only a small number of herring.

Why this type of year-class variation takes place remains unknown. The potential certainly exists for such vast population changes in consideration of the fact that a single fish may spawn thousands of eggs each year. But why one year-class can dominate to such a degree remains a puzzle. It could be a matter of ocean currents, food, predators, or cannibalism. Who knows?

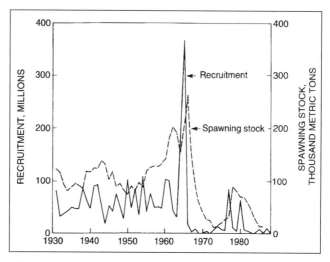

Figure 33. Estimates of Georges Bank haddock recruitment and spawning stock from 1930 to 1990. From Fogarty et al., 1992.

In any event, what has happened in more recent years off Norway is that overfishing and the lack of recruitment has led to the collapse of the fishing grounds.

Much the same has happened to the cod and haddock stocks on Georges Bank. As shown in the figure, there was a large recruitment to the haddock stock in 1965 which led to a correspondingly large spawning stock the following year. However, what has happened since is that lack of recruitment and overfishing have led to a collapse of the spawning stock with a partial recovery in the late 1970s to early 1980s and a final collapse to today.

At present both Georges Bank and the Grand Banks off Newfoundland, which has had a similar history, are both closed to fishing, leading to a severe downward plunge in the Newfoundland economy. It is uncertain when—or, for that matter, if—the cod and haddock stocks will be replenished sufficiently to allow these fishing grounds to be reopened. Some of the United States fishermen—working out of places like New Bedford—now fish further offshore and for such creatures as monk fish, which had not previously been considered to be of commercial value.

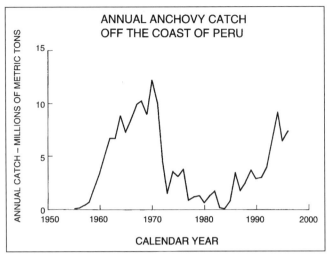

Figure 34. Peruvian anchovy catch from 1955 to 1996. Data from United Nations Food and Agriculture Organization.

The final figure shows the collapse of the anchovy fishing off Peru. During its peak years the anchovy catch amounted to almost 20% of the world's commercial fish catch. All the tiny fish were processed into fish meal for livestock feed—much the same as is done with the menhaden fish catch off the East Coast. During the years of maximum fish catch, the Peruvian government was alerted by marine biologists that the fishing was exceeding the recruitment of new anchovies to the population. In other words, they were overfishing and the total fish population was being diminished. These warnings were ignored.

Anchovies are plankton-eating fish. And, in turn, the plankton require abundant nutrients in the water for their photosynthesis. Off Peru these nutrients are supplied by coastal upwelling of nutrient rich, bottom water. However, every once in a while there are equatorial current changes that prevent the upwelling from occurring—a process known as El Niño. Such an El Niño occurred in 1972. The anchovy population collapsed and so did the Peruvian fishing industry. It has since recovered, at least partially, but it may be headed for another collapse.

Things are not much better for the top predatory fish—tuna,

shark, marlin, and swordfish—that are ubiquitous to all the world's oceans rather than to a specific fishing ground. They are all on the decline; and as of April 1999, the National Marine Fisheries Service has imposed new fishing restrictions on these migratory fish in the coastal waters from Maine to Texas. Swordfish, in particular, have been in sharp decline. Their populations have decreased dramatically as more and more juveniles are caught—a sure warning sign of overfishing any species. The average size of a swordfish is one-third of what it was thirty years ago.

In summary, the future for supplying the world's population with food is grim. We have pretty much reached the limits of exploiting arable land and fishing around the globe. The Green Revolution with the introduction of hybrid crops and improvements in irrigation and mechanization in the lesser developed countries will continue but it is doubtful that it will keep up with the population explosion. As the rural poor of the world move into cities, the cities compete for limited supplies of water, to the disadvantage of agricultural lands. The solution lies with population stabilization.

PART II
Tomorrow

CHAPTER 7

There is Much More to be Done and it Will Require the Concerted Efforts of All of Us

Introduction
What can be learned from the past as a guide to where attention should be placed in the future? We may expect a continuing preservation of the environment and curtailment of degradation at the local and regional level through the actions of concerned individuals and environmental groups. But there are two major problems which require action on a larger scale—global population growth and alternative energy sources. The first is a sociological problem and requires an International Initiative. The second is a scientific problem and requires at least a National Initiative. And these two problems we cover in the final two chapters—their causes, their history, and their solutions.

Past Successes and Failures
Each of the stories in the previous six chapters has a history and a lesson. They do tell us what we may expect in the future. So, let us look back at each of them in turn.

There is no more smog over London—at least, smog as originally defined. There is no more soft coal burning in the city. The Pea Soupers are a thing of the past. Most of the blackened buildings and statuary have been cleaned, though some of the statues were permanently damaged. To be sure, all of this came about because of a crisis—a situation which is common to many environmental stories. In this case, it was the Killer Smog of 1952.

Photochemical haze over Los Angeles still exists but there have been vast improvements in the efficiency of automobile engines—the chief contributor to the haze—and effective controls on noxious emissions in the form of catalytic converters. While the number of automobiles has quadrupled over the past 35 years in the greater Los Angeles area, the air pollution levels have dropped by 50%, representing an improvement by a factor of eight for emissions per automobile. And there are mandates for strict use of cleaner burning fuels, such as methanol, in Los Angeles by early in this century.

Nationwide the improvements have been equally dramatic. Over the past 25 years airborne levels of lead have declined 98%. Annual emissions of carbon monoxide are down 24%. And emissions of fine particulates have fallen 78%. Hardly pristine conditions, but not a bad record and one that will continue to improve through local, state, and federal regulations and watchdogs.

Acid rain remains a serious problem but here, also, there have been improvements. Impelled by public concern and legislation from Congress many of the electric utility companies in the Midwest have installed scrubbers to remove the noxious sulfur dioxide prior to flue gas emission. As a result, sulfur dioxide emissions have been reduced by 35% over the past 15 years while at the same time the use of coal has increased by 50%. This is certainly a substantial first step. Future improvements in retrofitting old plants with scrubbers as well as installing them in new plants may be expected.

The ozone layer problem—whatever its severity—has been solved. There is an international agreement to stabilize and ultimately eliminate production of CFCs globally. In retrospect, it was an easy problem to solve. There was an identifiable culprit, the CFCs, for which a nonpolluting substitute could be quite easily supplied by industry. And efforts toward resolution of the problem had strong support from both the scientific community and the public, the latter largely through the efforts of the media.

The health of Lake Erie has improved. Phosphorus is the limiting nutrient for photosynthesis, and consequent eutrification, in the lake. In the past a substantial portion of phosphorus input to the lake came from household detergents. With the use of phosphorus in detergents banned in 1970, the total phosphorus input to the lake has since decreased by 50%. The lake will never return to the pristine condition it enjoyed before the industrialization of its shores but the present improvement is a credit to all who have been concerned about the lake's degradation.

Similar improvements have taken place in the health of other bodies of water with the curtailment of chemical, industrial, and municipal waste discharges. Notable among these are the Hudson, Housatonic, and Connecticut Rivers in the northeastern portion of the country. But the same cannot be said for Chesapeake Bay. There the limiting nutrient for photosynthesis is nitrogen and the agricultural and municipal runoff continues unabated. The famous Chesapeake Bay oyster may soon be a thing of the past; their tolerance of adverse environmental conditions is limited. And other bodies of water remain polluted from waste discharges, prominently the section of the Mississippi River from Baton Rouge to New Orleans which is bumper-to-bumper with petrochemical and other related industries.

The Ogallala Aquifer is the largest known, underground supply of water in the world and it is being depleted for agricultural purposes at a rate greater than it is being recharged, a scenario which, if continued, spells disaster. The farmers of the High Plains region know this and have formed a set of Groundwater Management Districts consisting of a few to several adjacent counties. The GMDs have been effective in conserving Ogallala water and may eventually reach the goal set by GMD No. 4 in Kansas of reaching a zero depletion rate, i.e., withdrawing no more water from the aquifer each year than is being replaced by recharge.

The Cuyahoga River fire was an unusual, even one-of-a-kind, event although there are other potential areas, particularly in some of the more heavily industrialized regions of Europe. Since the fire in 1969, the river has been cleaned up. New sewage treatment plants and sewer lines have been installed in Cleveland; the steel industries along the river have installed water pollution facilities; and there has been the set aside of the Cuyahoga Valley National Recreation Area. In all, this is another example of an environmental crisis leading to a solution.

The Centralia mine fire and the evacuation of the town are events which should never have happened. The fire could have been contained at the outset in 1962 at a modest sum rather than the ultimate expenditure of $49 million some twenty-two years later. It is an example of the bungling of the bureaucratic system at all levels—local, state, and federal. We may anticipate further examples of bureaucratic mishandling of environmental problems from time to time but it is reasonable to expect that there will be no more Centralias.

The Dust Bowl, as with the Cuyahoga River fire and the Centralia mine fire, was a singular event. There will be future overextensions of agriculture to marginal lands with consequent economic hardship when adverse climatic conditions prevail but not on the scale of the Dust Bowl—at least in the United States. And prosperity has temporarily returned to the High Plains region with the mining of the Ogallala Aquifer.

The indiscriminate use of pesticides through aerial spraying is a thing of the past. Programs such as that to eradicate the fire ant in the Southern states and the gypsy moth in the Northeastern states were failures and did more harm than good. Now insecticides and herbicides are species specific and are thoroughly tested before being approved by the federal government for general use. Another environmental success story, thanks largely to Rachel Carson and her book *Silent Spring*.

Love Canal was an unusual example of toxicity in the environment. It is still difficult to imagine how an elementary school could have been built over a toxic waste dump but it did happen. The principal lesson from Love Canal is the effectiveness that can result from the action by citizens' groups in correcting a real, or perceived, wrong. They joined in the political process and brought about the closing of the school and the relocation of the nearby residents. And today the Environmental Protection Agency has a Superfund for cleaning up toxic waste sites, such as Love Canal, although, to date, more money seems to be spent in litigation than on actual clean up.

Times Beach was a disaster that never was. There was no need to buy out and evacuate the residents of the town and no need to subsequently incinerate the soil. It can only be considered as a knee jerk reaction by the Environmental Protection Agency and came at a low point in the affairs of that agency. At that time EPA was under considerable pressure for the way it was handling the $1.6 billion Superfund for toxic waste clean up for such places as Times Beach. The person in charge of the Superfund had been dismissed a few weeks prior to the announcement of the Times Beach buy out and the head of EPA was forced to resign a few weeks after the announcement. Bad decisions of this sort, made under pressure to attempt to alleviate an unrelated problem, will continue to be part of our political system.

Now, on to consider our decimation of other species. It is not a distinguished record. The rate of species extinction has risen as the global human population has risen. We were able to eliminate 50 million buffaloes in a twelve year period in the late 1800s and to kill some three to six billion passenger pigeons over a similar period of time also in the late 1800s.

Today, things are somewhat better but only that. There are no more decimations of the magnitude of the buffalo and passenger pigeon on land but there is continuing concern for the world's fishing grounds, many of which have been overfished and are now depleted in stock which was once abundant. There is concern for certain wildlife species in Africa and Asia, migratory birds in North America, and the numerous fauna and flora indigenous to the tropical rain forests. Fortunately, there are several environmental groups which continue to be active in restraining this species decimation.

Then, there are the examples of environmental degradation that remain with us and may never be corrected, or are beyond correction—such as Haiti, Aral Sea, Brazil, and Venice. There is no simple solution to the agricultural problems in Haiti. Once the nutrient rich soil on the mountain slopes is gone, it is gone basically forever. For the Aral Sea the recommendations from Arkady Levintanus are not likely to be effected. There is no longer a central government to adjudicate matters but rather a set of independent republics, some portions of which have benefitted from the river diversions and other portions of which have suffered from the disappearance of the Aral Sea.

For Brazil and Venice there has been some improvement. The rate of clear cutting in the tropical rain forest has declined; there are set asides for the rubber tappers; and there are refuges for faunal and floral preservation. In Venice there are no barrages to deter high water flooding, but groundwater withdrawal has been curtailed, and sewage treatment plants have been constructed so that the canals resemble somewhat less the open sewers that they were.

Finally, there are the potentially enormous problems that may result from global warming—both the direct effects of the warming, itself, and the indirect effects that it may trigger. Global warming is an environmental problem that cannot be ignored. The only solution lies in curtailing the burning of fossil fuels and the substitution of alternative energy sources which do not have harmful, by-product emissions for them.

In summary, the record shows that, by and large, when environmental problems have arisen and reached a crisis stage, we have reacted and solved them—most of the time. The concerned individuals and environmental groups and now, also, the Environmental Protection Agency and the corresponding state agencies remain in place and are active. We may be assured that future problems on a local and regional scale will be addressed—at least in the United States.

Global Environmental Problems

However, as we have mentioned throughout the book and will discuss in detail in the next two chapters, there are two problems which are beyond the mechanisms for problem-solving that are presently in place and which overwhelm—and drive—all other environmental concerns. They are global population growth and alternative energy sources.

The global population curve is as frightening an environmental statistic as one can put forward. The world's population has

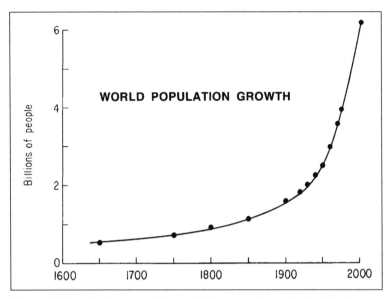

Figure 35. Population growth from 1650 to an estimated value for the year 2000 for the world. From Bogue, 1984.

risen from one billion in 1800 to an estimated six billion by the year 2000. And the scariest part of the graph is that the *rate* of population growth has continued to go up over the years in what looks like an ever-increasing fashion. When do we reach the point of overpopulation for this planet? We already have for much of the underdeveloped regions of the world, notably Africa and parts of Asia and South America. It is essential for human life on this planet that the global population become stabilized.

The second most frightening environmental statistic is given in the table on crude oil reserves and production. We live in an Oil Age. We are voracious consumers of a nonrenewable natural resource, the fossil fuels. Production worldwide in 1988 was 21.1 billion barrels and the ultimate reserves are 2,074 billion barrels. Simple arithmetic shows that, at the 1988 production rate, all of the reserves—all of the oil in the world—will be gone in 100 years. It will have taken only two centuries to exploit every bit of what it took the Earth a half-billion years to produce. And in fact it will probably be sooner, since the rate of production will almost

Region	Cumulative Production through 1988	Production in 1988	Identified Reserves	Ultimate Resources
North America	182.8	4.4	83.0	387
South America	57.9	1.4	43.8	142
Western Europe	15.7	1.4	26.9	69
Eastern Europe	6.8	0.1	2.0	11
Soviet Union	103.6	4.5	80.0	285
Africa	46.4	2.0	58.7	153
Middle East	160.2	5.1	584.8	867
Asia/Oceania	36.8	2.2	42.8	160
World	610.2	21.1	922.0	2,074

Table 4. World Crude Oil Production, Identified Reserves, and Ultimate Predicted Resources (billions of barrels). From Masters et al., 1990.

surely increase over the next decade or two. This is a most reckless spending of what could be thought of as national capital.

As it is essential that population become stabilized, it is equally essential that alternative and nonpolluting energy sources be developed. The solution of that problem serves as a solution to the problems of acid rain, smog, and potential global warming, all of which are directly related to the burning of fossil fuels.

Can these two problems be solved? If we put them on the front burner of our international and national concerns, the answer is yes.

CHAPTER 8

There are Two Major Environmental Problems that have to be Addressed—Global Population, A Sociological Problem,

Malthus Theorem

There would be few, if any, environmental problems if the world's population had remained at the 1800 level of one billion people, even considering the increased demand per capita for resources and energy over this same time period. Instead, with six billion human beings we are now pushing the limit of how many humans this planet can support.

A reasonable place to start in considering how large a population the Earth can sustain is Thomas Robert Malthus. Malthus was a nineteenth century, Anglican clergyman who had studied mathematics at Cambridge University. While it is a truism that population must always be kept to a level that can be supported by the food resources of the land, Malthus put this concept into quantitative terms in a series of treatises, beginning in 1798 with *An essay on the principle of population as it affects the future improvement of society.*

Malthus started with some simple mathematical considerations, comparing the hare of population increase to the tortoise of subsistence increase and concluding that the race would always go to the fleetest—population increase. For example, assume that the population in a given region doubles every generational reproductive cycle (25 years) producing a sequence of 1, 2, 4, 8, 16, 32, 64, 128, 256, 512, 1024...If the region were a small area like New Hampshire, with a 1980 population of about one million, the state's population would reach one billion in the next two and a half centuries. The notion of a billion people crammed into tiny New Hampshire is absurd, but it serves to illustrate Malthus's concept that population, like compound interest, increases in a geometric or exponential fashion and can easily get out of hand.

He framed what came to be called Malthus's dismal theorem.

1. Population is limited by the means of subsistence.

2. Population invariably increases where the means of subsistence increase, unless prevented by some very powerful and obvious checks.

3. These checks, and the checks which repress the superior power of population and keep its effects on a level with the means of subsistence, are all resolvable into moral restraint, vice, and misery.

For *moral restraint* substitute family planning. For *vice* substitute internecine struggles, ethnic cleansing, and racial and religious warfare. And for *misery* substitute starvation, droughts, floods, disease, and pandemics.

Malthus's ideas were controversial from the start, and for the most part were unpopular. He was looked upon wrongly as an advocate of war, pestilence, and other forms of misery and vice as checks on population. His later opinions, such as opposition to the English Poor Laws that established the right of the poor to relief, led people to see him as a toady of the rich. Furthermore, it turns out that his beliefs about population did not apply to the Europe of his day and were widely dismissed. Two major factors permitted the tortoise of subsistence to outstrip the hare of population: the Industrial Revolution with its great increase in both industrial and agricultural production, and the opening of the vast spaces of the Western Hemisphere.

A Malthusian exception took place in Ireland. With the introduction of the potato from the New World, the Irish population soared to more than eight million according to the census of 1841. But four years later, with the arrival of the potato blight, the crops failed, and again the following year. An estimated one million people died of starvation; many others emigrated to North America. Eventually, as a result of continued emigration and Malthusian "moral restraint" in the form of late marriages, the population stabilized at four million, or half the pre-blight figure.

Today, Malthusian vice and misery appear to be operating in the underdeveloped and overpopulated regions of Africa, Asia, and South America, where urban populations are growing exponentially along with urban crowding, poverty, and environmental degradation. The cyclones and typhoons that strike the densely populated lowlands of the Ganges delta region regularly take catastrophic numbers of lives. Wars, both large and small, seem constant; and AIDS and other diseases are on the march, devastating Africa in a manner reminiscent of the bubonic plagues of earlier centuries. Although global food production is currently sufficient to support a population of six billion, an adequate food distribution system is lacking and much of the food simply goes to waste. About 40 million people die each year of starvation and hunger related diseases and hundreds of millions suffer from malnutrition and undernutrition. In fact, it is estimated that more than one billion people, roughly 20% of the global population, live in poverty. The vast majority of these people live in the lesser developed countries—in Africa, Asia, and South America.

History of Population Growth

The global population growth rate obviously depends on the difference between the birth and death rates. (For a given country the growth rate depends on the difference between birth and death rates plus the difference between immigration and emigration rates.) But what may not be so obvious is that the death rate can be, and has been, as much a controlling factor as the birth

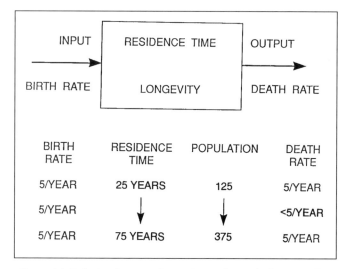

Figure 36. Relation between longevity and population growth.

rate. With more plentiful and varied food, with many infectious diseases under control, and with improved public health measures lowering infant and child mortality, the average life expectancy has increased from around twenty-five years in 1800 to seventy-five years today.

What are the effects of increased longevity on a population? For example, as illustrated in the accompanying diagram, assume an input to a community of five births per year. Under steady state conditions the output, or death rate, will also be five people per year. And with a residence time, or longevity, of twenty-five years, the population will be 125. Now, if the longevity is then changed to seventy-five years, the death rate will decrease until a new steady state is reached with a population of 375, being simply the product of the birth rate, five per year, and the new residence time, seventy-five years.

Thus, increased longevity can account for a substantial portion of the growth curve over the past two hundred years. In fact, it can account for a three-fold increase from one to three billion people, or 40% of the observed increase. This portion of the growth may be a one time effect; continued dramatic improvements in life expectancy are not likely.

Meanwhile, birth rates have actually declined, though not far enough to offset the increase in longevity. Birth rates typically do not adjust themselves as quickly as death rates, which are, of course, immediately affected by an increase in life span. In earlier times, large families were a necessity to offset high death rates and to create a large enough local labor force and therefore a civic virtue. Such attitudes take time to change, but the staggeringly high costs of child rearing and education are among the factors changing old ideas—particularly in the industrialized parts of the world.

The countries with high population growth rates are confined to the underdeveloped nations of the world, and this is where the global population problem lies. Annual growth rates of 2 to 3% are rampant. A growth rate of 3% may sound small but it implies a population doubling in a mere twenty-four years. Population increases in the same geometrical manner as principal in a savings account with compound interest.

For Europe, population stability has essentially been reached with very low to negative growth rates. For example, the present annual growth rate for Austria is 0.6%; that for England, 0.1%; and that for France, 0.3%. And for some countries there is a negative growth rate such as Bulgaria, -0.5%; Portugal, - 0.1%; and Romania, -0.2%.

In the United States things are improving. From 1975 to 1980 the annual growth rate was 1.1%; from 1985 to 1990, 1.0%; and from 1995 to 2000, a projected value of 0.8%. About half of the growth rate is related to immigration and about half to increased longevity.

The accompanying figure shows in some detail the changes in fertility in the United States from 1917 to 1994. Of particular note are the baby-boom years of the 1950s with high birth rates, the depression years of the 1930s with low birth rates, and the present birth rate of 2.0 births per woman which has been stabilized for the past twenty years. To the right in the figure is the term "replacement level". Replacement level fertility is the number a couple must bear to replace themselves. It is slightly higher than 2.0 to account for the fact that some female children die before reaching their reproductive years. If it were not for immigration and increased longevity, the United States population growth rate would be negative.

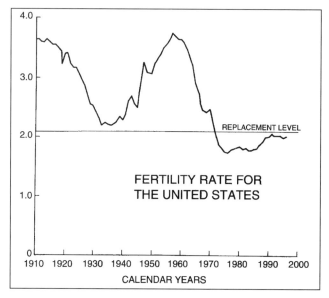

Figure 37. Fertility rate for the United States from 1910 to 1996. Data from the Bureau of Census.

For Africa and parts of South America and Asia population growth rates continue unabated. For Africa, as a whole, the average annual population growth is 2.6%. For some of the countries that continue to be prominent in the news because of their overpopulation problems we have growth rates of 2.8% for Burundi, 2.6% for Congo, 2.9% for Rwanda, 3.9% for Somalia, and 2.5% for Zambia. For South America things are less catastrophic but still there is a high growth rate of 1.5%. The problem areas in Asia are such places as Afghanistan with a growth rate of 5.3%; Iran, 2.2%; Iraq, 2.8%; and Pakistan, 2.7%.

As a counteractant to this population growth—a growth that is beyond what the land can provide in the form of subsistence—the Malthusian factors of misery and vice are much in evidence. Misery in the form of starvation and disease and vice in the form of warfare and ethnic killing.

Misery is evident everywhere. Its devastation can be followed from infants to children to adults.

Infant mortality rate is defined as the annual number of deaths of infants under one year of age per thousand live births.

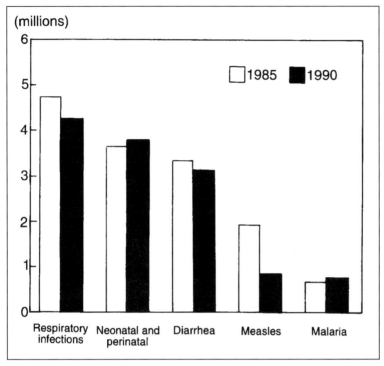

Figure 38. Estimated causes of death among children under age five in developing countries, 1985 and 1990. From World Resources Institute, 1992.

For the period from 1985 to 1990 the infant mortality rate for Africa was 103 as compared with 10 for North America, greater by a factor of ten. Put another way, one out of ten infants born in Africa dies in its first year. About half the deaths are related to malnutrition in the mother; most of the infant deaths in Africa may be considered preventable.

Children under the age of five comprise 37% of the deaths in the lesser developed countries as compared with 3% in North America and Europe. For 1990 the World Health Organization estimated that 12.9 million children died in the lesser developed countries. This amounts to 18% of the United States population. Or, to put it another way, it amounts to a number of children equivalent to the total population of New England dying each year. Infectious and parasitic diseases, most of which are preventable, are the main causes of death amounting to 9.8 out of the

12.9 million total. Pneumonia, influenza, diarrhea, whooping cough, measles, and malaria are the chief killers (see Figure 38).

Poverty, malnutrition, and outright starvation are the major killers in the adult population of the lesser developed countries. As mentioned previously, an estimated 40 million people die each year of starvation and hunger related diseases. There is simply not enough tillable land available to many of these populations to allow them all to survive.

Added to this is the devastation that may be forthcoming from widespread epidemics, or pandemics, particularly cholera and AIDS. Cholera can spread rapidly, particularly in overpopulated communities with poor sanitation and unsafe drinking water. There have been recent cholera outbreaks in South America and Africa but no pandemics, as yet. AIDS, on the other hand, is already a pandemic. Of the twelve million people worldwide whose blood is HIV-positive, eight million are in Africa. For Central Africa 10% of the population are affected and that number may grow appreciably over the next few decades.

The fact that there is an insufficient amount of land to support all the people in some parts of Africa and elsewhere has led to Mathus's final corrective measure—vice. There is not enough room for both my family and yours to survive. Your family, tribe, ethnic group, or whatever must be forced out or killed so that mine can exist. These are the bare facts in Bosnia, Rwanda, and Somalia. No amount of military aid or intervention will solve them.

Birth Control, Female Education

As one looks beyond the Malthusian effects of misery and vice to control the populations of Africa, Asia, and South America, two factors stand out.

One factor is obvious—accessibility to a variety of birth control techniques. The accompanying figure is a plot of average annual population change versus married couples using some form of contraception for all the countries in the world for which statistics are available. The lesser developed countries are crowded to the left in the diagram and the more developed countries to the right. The correlation is as expected. For Afghanistan, Cameroon, Mauritania, and Yemen the use of contraceptives is less than 2% and all these countries have an enormous popula-

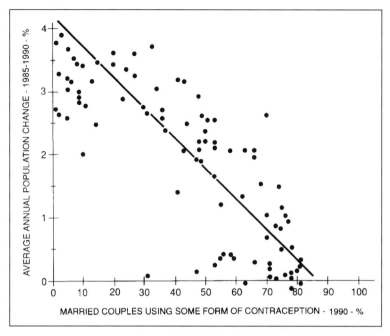

Figure 39. Average annual population change, 1985-1990, plotted against percentage of married couples using some form of contraception, 1990. Data from World Resources Institute, 1992.

tion growth problem as compared with 81% using contraceptives for Belgium, England, and France where there is no population growth problem.

The second factor that stands out as a population control may not seem so obvious—increased literacy and education of women. Education for women may be taken as a proxy for independence, self confidence, and self determination. On the accompanying graph the vertical scale is the same as that for the previous graph, average annual population change. The horizontal scale is the percentage of the female population for a given country that has completed primary school education. Again, the lesser developed countries are to the left in the diagram and the more developed countries to the right. Primary school education for females is less than 2% for Afghanistan, Guinea, Libya, and Mozambique as compared with 97% for the United States.

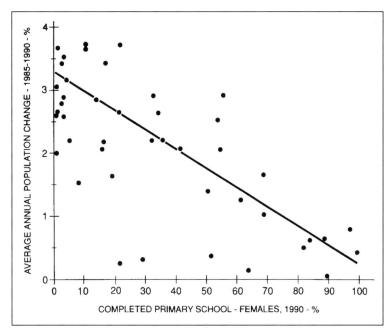

Figure 40. Average annual population change, 1985-1990, plotted against percentage of females having completed primary school. Data from World Resources Institute, 1992.

Moral and Religious Leadership

Before turning to what should be done and what part the United States should play, there is a third factor that could be very important in stabilizing populations in the lesser developed countries but, unfortunately, is sorely lacking. And that is moral and religious leadership—the Catholic church, the dominant religion in South America, and the Islamic faith, the dominant religion in Africa and Asia. The hierarchy in the Catholic church still has a hang up on the use of contraceptives. And the more fundamentalist sects within the Muslim faith treat women as chattel, as something lesser in the eyes of God than man.

The use of contraceptives has a long history in society, at least as far back as the Egyptian civilization. In the Catholic church the first, and also the most important, pronouncement on contraception came from the great theologian, St. Augustine, in the fifth century. It was simply that sexual intercourse was solely for the purpose of procreation. Not for pleasure, not for health, not for love in marriage. Now, Augustine would not be the first person that one would have selected for such a pronouncement. Prior to becoming a Catholic convert, Augustine had been somewhat of a reprobate. He had lived with a woman for eleven years from age eighteen to twenty-nine. He had a son by her in the first year of their relationship but had no intention of marrying her. He only gave up the relation at age twenty-nine when his domineering mother insisted that he become affianced to a more socially acceptable woman. He had no experience of love in marriage and the importance of sex to that relationship. To him sex was a sin as he came to consider his early eleven year relation.

Through the succeeding centuries the Church theologians ameliorated their position on sex in marriage, the use of contraceptives, and family planning. This all came to a peak at the time of Vatican II, the church council of 1964 under Pope John Paul XXIII. Apparently, most of the delegates to the council preferred a more understanding approach than that advocated by St. Augustine. Nothing came out of Vatican II itself but a special commission was set up to consider the matter. Its report under the imprimatur of the new Pope, Paul VI, came out in 1968 and is known as *Humanae vitae*.

To these deliberations many urged the Church to take a major leadership role in what was then and is still today our most important, long term problem—population control. One such appeal from forty-two British and European Nobel prize winners is quoted below. A similar letter from thirty-nine American, Argentinean, and Australian Nobel prize winners was also sent to the Vatican.

> His Holiness
> Pope Paul VI
> Vatican City
> Rome, Italy
>
> Your Holiness:
>
> We, the undersigned, Nobel laureates are conscious of the great responsibility borne by your Holiness in appraising and acting upon the advice offered by the

commission you have appointed to study the problems of population and fertility control. Because of the profound bearing of your decision on human welfare and happiness, now and for many years to come, we urge you to give due weight to the ever-growing opinion which contends:

That the uncontrolled growth of population is a major evil of present times;

That unwanted children are a source of unhappiness, privation, and distress; and

That parents should be able to exercise the right to have so far as possible, only the number of children which can be cared for and cherished.

Lord Adrian, Patrick Maynard Stuart Blackett, Max Born, Daniel Bovet, Lord Boyd-Orr, Sir Lawrence Bragg, Sir James Chadwick, Ernst Boris Chain, Sir Henry Hallett Dale, Sir Howard Florey, Werner Forssmann, Otto Hahn, Werner Heisenberg, Walter Rudolf Hess, Jaroslav Heyrovsky, Alan Lloyd Hodgkin, Dorothy Crowfoot Hodgkin, Andrew Fleming Huxley, Hans Daniel Jensen, Paul Karrer, John Cowdery Kendrew, Sir Hans Krebs, Halldor Laxness, Arthur John Porter Martin, Peter Brian Medawar, Giullo Natta, Max Ferdinand Perutz, Cecil Frank Powell, Salvatore Quasimodo, Sir C.V. Raman, Tadeus Reichstein, Sir Robert Robinson, Earl Russell, Frederick Sanger, Theodor Svedberg, Richard Lawrence Millington Synge, Arne Wilhelm Tiselius, Hugo Theorell, Sir George Paget Thomson, Lord Todd, Maurice Hugh Frederick Wilkins, and Karl Ziegler.

So, what was the end result? The Pope, in a central passage of *Humanae vitae*, declared:

Where conjugal intercourse is foreseen or carried out or brought to its natural results, there must be rejected every act whatsoever which, as an end to be accomplished or as a means to be applied, intends that procreation by impeded.

Back to square one.

What may the future bring? The Vatican is composed, largely, of elderly, celibate, Italian males. Now, there is nothing wrong with being elderly; the elderly have much wisdom. There is nothing wrong with being celibate, if that is one's natural choice. And there is certainly nothing wrong with being Italian or a male. But that combination has little to no experience with love and sex in marriage, family planning, and the very existence and continuity of families in the underdeveloped nations of the world. It is doubtful that there will be any change in the Vatican position in the forthcoming years.

There is hope that the national governments and local Catholic clergy in South America will take action on family planning and all that goes with it to control their burgeoning populations. Specifically to that point, in North America and Europe the Catholic laity pay little to no attention to the Vatican pronouncements on sex. For example, a survey taken during the period from 1982 to 1985 in the United States showed that 87% of Catholics as compared with 86% of Protestants, essentially the same number, approve of supplying birth control information to teenagers.

The more fundamentalist Islamic sects and the ones that dominate many of the regions of Africa and Asia with population growth problems are patriarchies. They are delineated by three strong characteristics. One, there is a culture versus nature dichotomy where man represents culture and woman represents nature, particularly the wilder and more uncontrolled aspects of nature. Man must subjugate woman. Two, there is a sexual division in labor in which woman's share is in childbearing and rearing and man's share is in production and power. The reproductive is secondary to the productive. And, three, men control the public domain and women are dominant in the private sphere with the public being the more important. In such a culture female education is unnecessary because it is not needed for women to perform their assigned functions.

Underlying all this is the notion that man represents the norm and the systematic and that woman represents the destructive and untamed. Islam views woman's sexuality as active, a threat to man, and therefore bad. Women's sexual powers left unchecked lead to destruction of the social order. Consequently, there is the practice of purdah, or seclusion of women from public observation among Muslims, and the barbaric coming of age practice in Central Africa of cutting out the female genitalia. Both actions aimed at desexing women.

It is very difficult to imagine that such a religion will ever come to accept equality of women and to consider the importance of female eduction and family planning. The following quotation from the February 16, 1996 issue of *The New York Times* illustrates what does happen when a fundamentalist sect takes over, in this case the Taliban in Afghanistan.

> Under Taliban decrees, women have been forbidden to work outside their homes, except in hospitals and clinics, and then only if they work exclusively with women and girls. Girls have been expelled from schools and colleges, and told that, for now at least, education is for boys only. Girls who were only months from finishing high school, or young women graduating from college, have been told their career dreams are over.
>
> Women wishing to go shopping in the bazaars, or to move anywhere outside their homes, must be accompanied by male kinfolk and wear the traditional burqua, a head-to-toe shroud with a netted slot over the eyes.
>
> The regime imposed by the Taliban, across a 600-mile stretch of territory from Herat in the west to the Pakistan border in the east, is one of such hostility to "modern" influences that the secretive Muslim clerics who lead the movement have ordered public "hangings" of television sets, video-cassette players and stereo systems. In Herat, Taliban fighters have gone from house to house pulling down satellite dishes and antennas, and confiscating books judged to be tainted by Western influence.

Family Planning

The factors of availability of birth control techniques and female education coalesce in the concept of *family planning*. Fortunately, some countries have included family planning as a national policy, China being the most notable—and also the most draconian. There, couples who pledge to have no more than one child are given extra food, larger pensions, better housing, free medical care, and salary bonuses; their child will be given free school tuition and preferential treatment in employment when he or she enters the job market. Couples who break their pledge lose all benefits. The result is that 83% of married couples are using modern contraception. And China, the world's most populous nation, now has an annual growth rate of only 0.9%.

Other countries, particularly in Southeast Asia, have followed more benevolent and conventional policies in family planning. For example, in Thailand 74% of couples now use some form of modern contraception, and that country's population change has been reduced from 2.4% in the period from 1975 to 1980 to 1.7% for the period from 1985 to 1990 to 0.8% for the period from 1995 to 2000. In South Korea, the figures are 79% using some form of contraception and population change reduced from 1.5% to 1.0% to 0.9%.

Thailand can deservedly stand as a model for other countries on how to achieve population stabilization. How has this come about? It has been a combination of several factors—each of which is important. The *federal government* has initiated a program in family planning; at the national level uncontrolled population growth has been recognized as a crucial problem. The program has the support of the *religious leaders*; 95% of Thais are Buddhist. The *populace* has been open to new ideas in family planning. There are *nonprofit family planning organizations* which are active and have the support of the government, particularly the Population and Community Development Association. And, then, as in many other environmental successes, there has been the *individual*: in this case Mechai Viravidaiya, a former government economist and a master at public relations.

Although it has a long way to go to achieve population stabilization, Bangladesh has added another dimension—*economic independence for women.* It is an innovative program introduced by the Grameen Bank. It consists of "microcredit" or small business loans to women. To date, the bank has loaned more than two billion dollars in small amounts to more than two million destitute women. Their pledge to the bank is that "we shall plan to keep our families small, we shall keep from the curse of dowry, and we shall not practice child marriage."

International Initiative

Despite the havoc engendered by the Malthusian adjusters of misery and vice, much of the world still has population growth rates of 2 to 3%. The problem areas include all of Africa, the Near East, Western South America, and Central America.

What has to be done is pretty obvious. It has been stated over and again by those concerned. Population growth has to be recognized as an international problem. In fact, it has to be recognized as our *most important international problem.* Everything else follows from that recognition.

The United Nations does recognize the importance of the global population problem. This was spelled out in its "Amsterdam declaration on a better life for future generations" of 1989. The preamble to that document states:

> — *Emphasize* our responsibilities towards future generations, in particular in the field of population, where the actions and decisions of one generation determine to a significant degree the demographic make-up of future generations and by implication the nature of the world and society in which the millions of people not yet born will live;
>
> — *Acknowledge* that population, resources and environment are inextricably linked and *stress* our commitment to bringing about a sustainable relationship between human numbers, resources and development;
>
> — *Express concern* that the continued rapid growth in world population, especially in the developing world, the processes of uncontrolled

migration and urbanization and the increasing degradation of environment everywhere threaten to darken our vision of the world we will leave for posterity in the twenty-first century;

— *Recognize* that women are at the centre of the development process and that the improvements of their status and the extent to which they are free to make decisions affecting their lives and that of their families will be crucial in determining future population growth rates.

— *Further recognize* that the principal aim of social, economic and cultural development, of which population policies and programmes are integral parts, is to improve the quality of life of the people.

The document goes on to state for its first three priorities and approaches that they should be aware of and sensitive to:

— The effects of education on demographic behaviour and the critical importance for development of increasing female literacy and achieving universal enrollment of girls in primary school by the year 2000;

— The need to raise the social and economic value of girl children in the family, community and national development;

— The need to increase women's participation in decision-making and management of population policies and programmes and special programmes for economic development of women, with the aim of achieving equality of representation.

The Amsterdam declaration calls on a funding level of $9 billion a year by the year 2000, a doubling of the funding level for 1989. The declaration and most population experts consider that this would be adequate to offer every woman in the world access to family planning services. That total dollar figure amounts to a commitment of a mere 4% of the foreign aid budgets from the more developed to the lesser developed nations. The portion for the United States would be $500 million, an amount of 1/20 of 1% of the U.S. national budget. That is a small sum of money to address our most critical environmental problem. The Amsterdam declaration was signed by seventy-nine countries including the United States; it is time to see it effected.

On the national level foreign aid from the United States to the lesser developed nations should be predicated on their having national programs in hand to stabilize their populations. Otherwise, the aid, including that for continuing the Green Revolution, has little meaning; poverty and starvation will continue to grow and reign.

Before investing in dams, or roads, or elaborate technical apparatuses of either peace or war, developing countries can help themselves most by investing in education and family planning. An encouraging sign is that a number of countries including the Muslim countries of Bangladesh, Indonesia, and Iran have instituted in recent years as a national policy a family planning infrastructure. There is only a small—but disproportionally vocal—constituency in the United States that would object to the federal government and various international development agencies and banks tying their aid and trade policies to female education and family planning. Such a group can be overruled when people recognize that old doctrines are at war with the inexorable course of reality.

CHAPTER 9

...And Alternative Energy Sources, A Scientific Problem.

Natural Resources Depletion

Our concern here is with the necessity to develop alternative and nonpolluting energy sources. It is a concern with substituting *renewable* energy sources for *nonrenewable* energy sources, i.e., the fossil fuels.

If you substitute in Malthus's dismal theorem the phrase "nonrenewable natural resources" for "subsistence", the consequences are even more dire. After all, most food stocks are replenished seasonally, but such items as fossil fuels and minerals are not. Once they have been used up, that is that; they simply are no longer available. And the graphs of the consumption of energy and mineral resources look very much like the growth curves of population. While there will always be land enough to grow food for *some* people, when fossil fuels and certain minerals are gone, no one will have them.

In fact, we discriminate the major eras of humankind's time on the planet by what we have extracted from the Earth beneath our feet. The Stone Age extended from prehistoric times to 3,000

POPULATION — FOOD — RESOURCE DEPLETION — POLLUTION

RESOURCE DEPLETION = $\sum_{\text{WORLD}}$ (CONSUMPTION PER CAPITA) x (POPULATION)

UNITED STATES — 180 x 10^6 BTU/CAPITA/YEAR

NIGERIA — 10 x 10^6 BTU/CAPITA/YEAR

Figure 41. *Population, food, resource depletion, and pollution.*

B.C., and beyond, locally. Next was the Bronze Age, bronze being an alloy of copper and tin, from 3,000 B.C. to 1,000 B.C. Then, the Iron Age from 1,000 B.C. to 100 A.D. or the present, depending on one's choice. One could say that we now live in the Oil Age, which began late in the nineteenth century during the "romantic period" of the extraction industries when engineers and entrepreneurs eagerly roamed the Earth's surface, like raiders of the lost arc, looking for mineral and oil riches.

The heady days of mineral and petroleum exploration are over, times when the rugged individualists prospected, made fortunes, lost them, then moved on, opening up remote areas to settlement, times like the California gold rush and the Texas oil boom. Whether the mythology corresponds with the reality of those times or not, the reality today is sobering. You need to drill far deeper or further offshore for oil, and only huge corporations or conglomerates of small firms can play the game. Most of the major, near surface, easily mineable ore deposits of minerals have been found and exploited, and production comes from increasingly great depths and from lower grade deposits. We are squeezing the last juices out of the fruit and, as a result, they are getting more expensive.

There is an obvious interconnection between population, food, resource depletion, and pollution. The greater the population, the greater the demand for food and natural resources. And the greater the resource depletion, particularly that of the fossil fuels, the greater the global pollution. To compound the resource depletion problem is the fact that not only has the global population increased but the consumption of natural resources per person has also increased—the total resource depletion being the

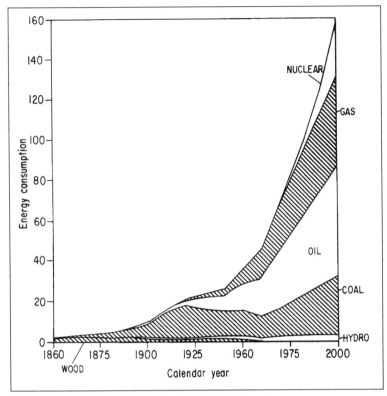

Figure 42. United States energy consumption from 1850 to projected level in 2000. Energy scale is in 10^{15} BTU (British thermal units) per year. From Starr, 1971.

product of the two. Natural resource depletion is as severe a global problem as sustainable agriculture and population growth. Whereas the global population growth problem lies with the lesser developed countries, the resource depletion problem lies with the more developed countries because of their much higher energy consumption per capita (see Figure 41).

Specifically, the accompanying figure shows how energy consumption has increased as well as changed in type over the past 150 years in the United States. In 1850, wood supplied 90% of all the energy used in a young United States—a renewable resource, by the way, so long as decent forestry practices are used. By 1910, coal supplied 75% of American energy. And by 1970, oil and gas provided

75%. Today, the contribution of the three fossil fuels—coal, oil, and gas—account for 85% of United States energy consumption with the contribution from nuclear reactors in generating electricity becoming increasingly more important.

It is clear from the graph that U.S. energy use has risen exponentially. What the graph does not make clear is that while the population in the United States will have increased twelve times from 1850 to 2000, the increase in energy consumption is three times that. It will have grown thirty-six times over the same period. Worldwide, energy consumption appears to be rising at a rate of about 3% annually, the major portion of it in fossil fuels. And this is occurring just at the point when the by-products are especially unwanted—notably, acid rain, atmospheric pollution, and probable greenhouse warming.

A pertinent question to ask is how long this profligate mining of fossil fuels and other nonrenewable resources can continue. A look at the table in Chapter 8 on ultimate reserves of crude oil, the major energy source of the United States and the rest of the industrialized world, is far from reassuring. As stated in that chapter, at present consumption rates all of the oil will be gone in a hundred years. If the annual growth rate of 3% in energy consumption is also factored in, all of the oil will be gone in fifty years.

Another uncomfortable and well known fact is plainly shown in the table. If we continue to rely on oil, we will have to rely as well on the political and international responsibility of the Middle East—one of the most politically volatile places on Earth. For there, in the crucible of three of the world's major religions, are also found 63% of the identified world reserves of oil and 43% of the ultimate predicted reserves.

The corresponding figures for natural gas give little more encouragement. At current rates of use, the world supply will last for a hundred years or slightly longer. For coal, supplies will continue for a few hundred years, perhaps as long as five hundred years. Of the three, coal is the dirtiest; the more that is left in the ground, the better.

Then, there are the immense deposits of oil shales in the Western United States and tar sands in Western Canada and the Middle East. At present, their high extraction and refinement costs prohibit them from being considered as a reserve. However, if the price of gasoline rises high enough in the first half of this century, then these deposits could come into play. Their conversion to liquid petroleum would extend the fossil fuel limits for a few hundred years more. But this extension will cost dearly; their extraction will mainly be by surficial, strip mining techniques over areas the size of small countries, and the environmental consequences of such strip mining will be immense.

With these grim figures in mind, one might reasonably ask why the oil problem has not risen to the forefront of national and international concerns. To be sure, there have been efforts—greater fuel efficiency for automobiles, improvements in mass transportation, and increased utilization of nuclear, solar, wind, and geothermal power. But these have been driven as much by concerns about pollution as by resource considerations. With the Arab oil embargo of the late 1970s, an effort was initiated by the federal government to develop synfuels, i.e., the conversion of plants, a renewable resource, into clean burning fuels. But these latter efforts have diminished in recent years.

There are two reasons why the natural resource problem, itself, has not received the attention it deserves. First, cries of wolf-wolf date back to the 1950s, and beyond, predicting that we were running out of oil. But vast new deposits were soon found and exploited—Alaska, offshore California, the Gulf of Mexico, the North Sea, and Nigeria, for example—and the wolf didn't show up at the door. We now have reliable figures for the volumes of the sedimentary basins of the world and how much oil they may contain. Today, at least for liquid petroleum, the wolf is at the door. The party will end in the foreseeable future.

The second reason is shown in the graph of the variation in the price of gasoline over the past half century. The important feature is the price of gasoline as adjusted for inflation. Except for the Arab oil embargo period of the 1970s, the price of gasoline has actually gone down over the past forty-five years. So, where is the problem? The problem is that there has seemingly been no problem, but the Arab oil embargo price jump up to $2.25 per gallon as adjusted for inflation is only a precursor to what will happen in this century. As the supplies diminish and oil becomes more valuable, prices will have to rise, and keep rising.

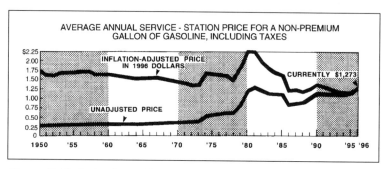

Figure 43. Variation in gasoline price over the past twenty years. From The New York Times, May 3, 1996.

Research and Development

We need to recognize now the importance of the global energy problem; and in the United States, we need a national initiative to solve it. We need a full-scale program to develop alternative energy sources—to diminish our dependence on the nonrenewable energy sources of the fossil fuels and replace them with renewable and nonpolluting energy sources.

It is an initiative which will require both research and development. By research we generally mean scientific studies to better understand the nature of things and the physical, chemical, and biological processes involved. By development we generally mean engineering innovations from these scientific findings for the benefit of humanity. For all this, the federal government must play an important role in supplying the needed funding. Technology addressed at environmental problems is not to the advantage of one specific industry or interest group but rather to the citizenry at large—to the benefit of all of us. It is our right to demand that our tax dollars be used for our benefit.

As a nation, we have been very successful in such research and development initiatives in the past. During World War II technology provided us with radar for the detection of aircraft, advanced sonar for the detection of submarines, and the atomic bomb, among other developments. The Apollo program, a concentrated and dedicated effort by NASA, led to the landing of a human on the Moon. The Cold War led to further military developments with spinoffs for civilian use, such as the supersonic transport.

Messerschmitt 163 (Komet). From Munson, 1969.

Now that the Cold War is over, it is time to put environmental problems in the forefront of our considerations as they are the most important long term problem we face. If we recognize them as such, we can address them in the same successful manner as we did World War II, the Apollo program, and the Cold War.

We live in a technological, as well as an oil age. We simply take that for granted and often fail to appreciate all the technological improvements that have taken place over the past thirty to fifty years, and equally fail to appreciate all the technological improvements that will occur in the near future. We take for granted the ubiquitous TV; computers, in general, and personal computers, in particular; open-heart surgery, kidney replacement…the list goes on.

To illustrate the importance of technology and also the unpredictability at the outset of some projects of the ultimate developments that can result from a given project, consider the example of the supersonic transport, or SST. It began with the small airplane shown in the accompanying photograph, the first mass produced jet aircraft, back in 1945. It was a German fighter plane known as the Komet or from its manufacturers' designation as the Messerschmitt 163.

The Komet had some unusual features for an airplane of that era. It could climb to an altitude of 30,000 feet in 2-1/2 minutes, an unheard of accomplishment. The maximum speed at altitude was 600 miles per hour, 150 to 200 miles an hour faster than conventional fighter aircraft of World War II. The Komet was also the first introduction of the swept back, delta wing with no tail assemblage.

But it also had some serious deficiencies. Having reached its altitude, it had a cruising time of only 10 minutes—that's right, a mere 10 minutes. It was used strictly as a home defense fighter, sent up to attack allied bombers as they came over a given German industrialized region. After using up its fuel, the Komet descended and landed as a glider. (It had wheels for take off but these were discarded as soon as the plane was airborne.) Another serious deficiency was the dangerous fuel that was used, a mixture of liquid hydrogen and hydrogen peroxide. About a third of the mishaps to the Komets came from explosions on fueling the plane, or from hard landings with the explosion of the remaining fuel in the plane.

Now, if you project yourself back to 1945 with the knowledge in hand of the Komet and its serious deficiencies, how many would have been intrepid enough to predict, at that time, that only 25 years later, in 1970, we would have the supersonic transport with regularly scheduled flights from London or Paris to New York or Washington, capacity of 300 passengers, speed of 1,800 miles per hour, and a crossing time of 4 hours. In other words, a flight from Paris to New York arrives two hours earlier by the clock than the time it departed. And all this using conventional fuels.

We do not think that back in 1945 there would have been anyone but the most wild-eyed airplane designer who would have predicted such a scenario. And that is an important lesson in itself. We often cannot predict what technological advances will be made or how quickly they will be made. But we can be assured that they *will occur*—particularly if we put our collective minds to demanding them.

The advance from the Komet to the SST did not occur accidentally. It was a well planned and well carried out product of the Cold War. Following World War II, federal financing through the Defense Department led to the development of jet aircraft and

Concorde (supersonic transport). From Knight, 1976.

then supersonic aircraft, in particular. The results from the military development spread over into civilian aircraft. But the important ingredient was the *impetus* from the federal government, which had the research and development funding to back it up.

Alternative Energy Sources

So, what are the technological problems to be addressed and solved? There are several avenues for research and development, some of which will be successful and some not, but all of which should be pursued. The rewards of technology are there if we have the will to seek them.

The greatest source of renewable energy available to us is that radiated from the Sun. It is the energy source for photosynthesis, the energy source for life on Earth; and it is underutilized. There are experimental cars driven by solar energy. How much further can we go? Solar energy for heating already exists to a modest degree. Solar heating for electrical power generation is a reachable goal needing more research and development.

Wind power generation is an indirect result of solar energy, and experimental wind-mill farms already exist. At what stage of development and/or electricity price rise does wind power become economic. We already utilize the Earth's latent heat in the hot springs regions of California, Iceland, and New Zealand for heating and electrical power generating purposes. The challenge lies in those regions where there are hot, dry rocks at depth but no available circulating water.

Biomass energy conversion is another indirect result of solar radiation, and one of the more obvious and more achievable alternate energy sources are synfuels. The distillation of terrestrial and aquatic plants to methanol, ethanol, and alcohols as a replacement for gasoline in the automobile works, but crop selection, scaling, process development and the like are needed.

Nuclear energy is a nonrenewable but nonpolluting energy source. In the years immediately following World War II, nuclear energy was seen to be a panacea to the ever-growing need in the world for electricity. It would be, it was said, too cheap to meter. It has not yet become anything of the sort, nor will it ever. The present generation of reactors are extraordinarily fuel inefficient. What is immediately needed is the next generation of reactors, the so-called breeder reactors, which are about a hundred times more efficient. In the future, there remains the indistinct possibility of fusion reactors which would be a giant step upward in energy generation.

Meanwhile, the need for enhanced research into understanding the effects of our continued burning of the fossil fuels is crucially important, involving atmospheric pollution and acid rain and their environmental effects, the carbon dioxide cycle in the atmosphere, the greenhouse process, and the possible effects of global warming on the environment and on its fauna and flora including humanity.

Others could add to the list. The point is that it is not difficult to assemble a list of research and development directions on alternative energy sources.

National Initiative

Where do the research and development funds come from? The Cold War was by far and away our major research and development expenditure over the past several years. Now that the Cold War is over, it is time to recognize that the environment demands the same kind of attention for our continued well being and benefit—not just because it would be a nice thing to do but because it is necessary.

So, let us start by looking at the federal budget figures, in general, and then go on to look at the research and development figures, in particular.

The unit of money that will be used in this comparison is billions of dollars. On a personal basis it is difficult to conceive of what the expenditure of a billion dollars per year means. A million dollars a year, yes; even ten million a year; but a billion is beyond normal, every-day comprehension. On the other hand, the federal budget is in excess of a trillion dollars. Thus, a hundred billion dollars represents 10% of the federal budget and ten billion dollars represents only 1% of the budget. A billion dollars is a meager 1/10 of 1% of the federal budget—an insignificant amount, lost in the background noise of Federal budgeting.

The figures in the first table are the budget estimates for 1999 for all the major departments and for the minor departments of particular interest here. They have been listed in decreasing order of funding. (The subheadings within each department are a few items of interest; their sum is not the same as the total budget for that department.)

The most striking thing about the budget figures is that the first three entries—Social Security, Health and Human Services, and Defense—dominate the picture. Entitlement and Defense amount to over a trillion dollars. Everything else, with the single exception

Department or Agency	Billions of Dollars
Social Security Administration	
Total	429.9
Department of Health and Human Services	
National Institutes of Health	14.1
Health Care Programs	229.2
Total	380.9
Department of Defense	
Missile procurement, Army	1.2
Aircraft procurement, Navy	6.2
Weapons procurement, Navy	1.3
Shipbuilding, Navy	7.9
Aircraft procurement, Air Force	7.7
Missile procurement, Air Force	2.3
Research, development, test, and evaluation	36.7
Total	258.5
Department of Agriculture	
Forest Service	3.2
Total	57.4
Department of Energy	
Weapons activities	4.5
Defense environmental restoration and waste management	4.2
Total	18.3
Department of the Interior	
Geological Survey	0.8
Fish and Wildlife Service	1.1
National Park Service	1.9
Total	9.6
Department of Commerce	
National Oceanic and Atmospheric Administration	2.1
Total	4.9
National Aeronautics and Space Administration	
Space Station	2.3
Total	13.4
Environmental Protection Agency	
Hazardous substance superfund	2.2
Total	5.8
National Science Foundation	
Total	2.6

Table 5. 1999 Budget Estimates.

Department or Agency	Billions of Dollars	Percent
Defense	36.7	48.0%
Health and Human Services	14.1	18.4
National Aeronautics and Space Administration	9.6	12.6
Energy	7.0	9.2
National Science Foundation	2.6	3.4
Agriculture	1.6	2.1
Commerce	0.8	1.0
Interior	0.6	0.8
Transportation	0.9	1.2
Environmental Protection Agency	0.6	0.8
Veterans Affairs	0.7	0.9
All other	1.2	1.6
Total	76.4	100.0

Table 6. 1999 Budget Estimates— Research and Development.

of Agriculture, is quite small by comparison. On the other hand, the sum total of funding for the agencies with a long-standing obligation to preserve and protect our environment, specifically, the Forest Service, Geological Survey, Fish and Wildlife Service, National Park Service, and National Oceanic and Atmospheric Administration, amounts to less than 1% of the Federal budget. The addition of the Environmental Protection Agency in the 1970s adds another fraction of a percent. The environment, in other words, does not rate very high in the federal budgeting process. A fair amount of lip service is given to the environment during political campaigning—in the same category as God and motherhood—but little to no action after the election.

The figures in the second table are the 1999 budget estimates for research and development. It is dominated by Defense research and development which amounts to nearly 50% of the total budget. Looking down the list, the first four entries—Defense, National Institutes of Health (NIH), National Aeronautics and Space Administration (NASA), and Energy—comprise 88% of the budget. The Department of Energy (DOE) figure is largely for nuclear weapons research, an outgrowth from the old Atomic Energy Commission which is now part of DOE.

All the other entries including the National Science Foundation and the Environmental Protection Agency make up the remaining 12%.

The 14.1 billion figure for NIH and the 9.6 billion for NASA are related in substantial part to the sophisticated job that both agencies do in getting their message of research results across to the public and, also, to Congress. They make the effort to go over their research results with the public in a clear and understandable—even, at times, melodramatic—way. From NIH we are continually made aware of advances in research on AIDS, Alzheimer's disease, cancer, and heart surgery. From NASA we have had in the past the Apollo program with the landing of a human on the Moon and now the possible existence of life at one time on Mars. Environmental research has not received the same attention.

Next down the list is the National Science Foundation (NSF). Although small in budget amount, it is a particularly important part of our national research picture. A great deal of our scientific talent lies in academic institutions, and a substantial portion of our achievements and advances in basic research come from this sector. NSF has been important in funding academic research. If, as is argued here, increased emphasis on environmental research is warranted and should be a priority item, then the academic community should be brought in as a major player. NSF's budget should be increased by a billion dollars—a tiny amount in the overall budgeting—to assist in funding such research.

Finally, let us look at how the federal expenditures for research and development have varied over the recent past. The accompanying graph is a plot of such expenditures, as adjusted to 1992 dollars, over the past fifty years. During the 1950s the total federal expenditures for research and development increased gradually year by year with the bulk of the expenditures being in defense related research. Then, in 1959 with the advent of the Cold War defense research took a big jump to sustained levels of 22 to 32 billion dollars a year for the 1960s and 1970s. Nondefense research expenditures were soon to follow, reaching levels comparable with defense research expenditures. Then again, in the late 1970s defense research made another big jump to 45 billion dollars during the 1980s. With the demise of the Soviet Union,

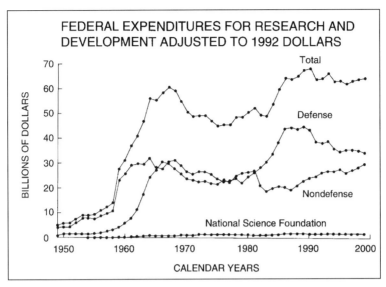

Figure 44. Federal expenditures for research and development as adjusted to 1992 dollars. Data from Office of Management and Budget, 1998b.

Cold War research expenditures have since decreased by 10 billion to the present level. During this same period nondefense research expenditures declined and then gradually recovered to their 1980 level. By comparison with the enormous increase in total research and development from a level of 5 billion, as expressed in 1992 dollars, in 1949 to a level of 65 billion in 1999 and the specific increases for the NIH and the NASA from zero in 1949 to 14.1 and 9.6 billion in 1999, respectively, the increase in funding for the National Science Foundation since its inception in 1954 has been quite modest.

What kind of changes are in order? Not an increase in total research and development expenditures but a redress with emphasis on nondefense research as compared with defense research. A gradual increase in funding for environmental science and technology of up to 10 to 20 billion dollars a year and a corresponding decrease in Cold War science and technology and procurement, which are no longer needed.

This redress does not refer to any military reductions related to carrying out limited wars such as the Persian Gulf or police actions such as Somalia, Bosnia, and Kosovo. It refers only to research, development, and procurement related to the Cold War. These are the large scale and expensive items. In particular, they include nuclear weapons and their delivery systems—intercontinental ballistic missiles, nuclear missile carrying submarines, and nuclear bomb carrying aircraft.

For the 1999 budget there is funding for procurement of an additional attack submarine at over two billion dollars and an advanced destroyer at over three billion dollars. Is there really a need for these additional weapons aimed, in part, at attacking and destroying alien ballistic missile carrying submarines. Russia, the only possible threat in this regard, is in the process of decommissioning its ballistic missile submarine fleet. Since 1992, they have taken out of service 25 such submarines, leaving an operational fleet of 20 ballistic missile submarines with 4 more scheduled to be retired soon. And, anyway, we already have 65 attack submarines and 50 destroyers in operation. At a time when Russia is reducing its threatening fleet, are we justified in increasing our deterrent force?

For the 2000 budget the Defense Department has proposed a funding of *350 billion dollars spread over two decades* for the new F-22 superfighter and two other similar aircraft or, on average, 17.5 billion a year. The Soviet Union no longer exists and there is no competitor on its scale in high technology weapon capability. We and our allies have supreme air control, as demonstrated in the Kosovo police action. Is such an enormous expenditure for a next generation of weaponry justified?

It comes down—always—to a matter of priorities. Do we need more Cold War weapon systems, research, development, testing, evaluation, and procurement or do we now need to pay more attention to the several environmental problems that have to be addressed?

So, there you have it. An international effort to curb global population growth and a national program to develop alternative and nonpolluting energy sources. Both can be accomplished if it is our will to do so.

REFERENCES

PART I
Chapter 1
National Parks, Forest Preserves and Wildlife Refuges—our national heritage

Harris, A.G., 1975, Geology of National Parks, Kendall/Hunt Publishing, Dubuque, 299 pp.

Ise, J., 1961, Our National Park policy, The Johns Hopkins Press, Baltimore, 701 pp.

Lindt, D.L., 1979, Ding: the life of Jay Northwood Darling, Iowa State University Press, Ames, 202 pp.

McGeary, M.N., 1960, Gifford Pinchot: forester, politician, Princeton University Press, Princeton, 481 pp.

Runte, A., 1987, National Parks: the American experience (second edition), University of Nebraska Press, Lincoln, 335 pp.

Turner, F., 1985, Rediscovering America: John Muir in his time and ours, Viking, New York, 417 pp.

Haiti—deforestation

Cobb, C.E, 1987, Haiti against all odds, National Geographic, 172, 645-675.

Heinl, R.B. and N.C. Heinl, 1978, Written in blood: the story of Haitian people, 1492-1971, Houghton Mifflin, Boston, 785 pp.

Kurlansky, M., 1988, Haiti's environment teeters on the edge, International Wildlife, 88, 2, 34-38.

Lewis, L.A. and W.J. Coffey, 1985, The continuing deforestation of Haiti, Ambio, 14, 158-160.
MacLean, F., 1987, Henry Christophe, legendary King of Haiti, Smithsonian, 18, 7, 160-173.
MacLean, F., 1993, When the marines went to Haiti, Smithsonian, 23, 10, 44-55.
Riding, A., 1974, Haiti losing fight against erosion, The New York Times, 23 June 1974, 15.

Brazil—tropical rain forests
Hoge, W., 1980, Development is eating up the world's rain forests, The New York Times, 31 August, 16E.
Medina, J.T., 1934, The discovery of the Amazon, American Geographical Society, New York, 467 pp.
Nepstad, D.C., C. Uhl and E.A.S. Serrão, 1991, Recuperation of a degraded Amazonian landscape: forest recovery and agricultural restoration, Ambio, 20, 248-255.
Parfit, M., 1989, Whose hands will shape the future of the Amazon's green mansions, Smithsonian, 20, 8, 58-75.
Revkin, A., 1990, The burning season, Houghton Mifflin, Boston, 317 pp.
Salati, E. and P.B. Vose, 1983, Depletion of tropical rain forests, Ambio, 12, 67-71.

Chapter 2
Venice—Disappearance of a Venerable City
Fay, S. and P. Knightly, 1976, The death of Venice, Praeger, New York, 190 pp.
Hamblin, D.J., 1977, Maladies of Venice: decay, delay and that old sinking feeling, Smithsonian, 8, 8, 40-53.
Hofman, P., 1973, Foul-air masks ordered for Venice, The New York Times, 8 January, 1 and 5.
Judge, J., 1972, Venice fights for life, National Geographic, 142, 591-626.
Maciolek, C.R., 1993, Operation in recovery, Sky, 22, 4, 32-40.
Morris, J., 1974, The world of Venice, Harcourt Brace Jovanovich, New York, 315 pp.

Aral Sea—Requiem for an Inland Sea
Brown, L.R., 1991, The Aral Sea: going, going,..., World Watch, 41, 20-27.

Ellis, W.S., 1990, A Soviet sea lies dying, National Geographic, 177, 2, 73-93.
Kotlyakov, V.M., 1991, The Aral Sea basin: a critical environment zone, Environment, 33, 1, 4-9 and 36-38.
Levintanus, A., 1992, Saving the Aral Sea, Environmentalist, 12, 85-91.
Micklin, P.P., 1988, Desiccation of the Aral Sea: a water management disaster in the Soviet Union, Science, 241, 1170-1176.
Micklin, P.P., 1993, The shrinking Aral Sea, Geotimes, April, 14-18.
Precoda, N., 1991, Requiem for the Aral Sea, Ambio, 20, 109-114.
Reznichenko, G., 1988, The Aral: who is to blame and what is to be done, Moscow News, 42, 9.
Sinor, D., 1969, Inner Asia, Indiana University Press, Bloomington, 261 pp.
Wheeler, G., 1964, The modern history of Soviet Central Asia, Frederick A. Praeger, New York, 272 pp.

Dust Bowl—Desertification

Bonnifield, P., 1979, The Dust Bowl: men, dirt, and depression, University of New Mexico Press, Albuquerque, 232 pp.
Hurt, R.D., 1981, The Dust Bowl: an agricultural and social history, Nelson-Hall, Chicago, 214 pp.
King, S.S., 1976, Drought perils wheat in the Dust Bowl, The New York Times, 29 February, 1 and 23.
Kromm, D.E., 1992, Low water in the American High Plains, World & I, 7, 312-319.
Lockeretz, W., 1978, The lessons of the Dust Bowl, American Scientist, 66, 560-569.
Parfit, M., 1989, You could see it a-comin, Smithsonian, 20, 3, 44-57.
Risser, J., 1981, A renewed threat of soil erosion: it's worse than the Dust Bowl, Smithsonian, 11, 12, 120-131.
Svobida, L., 1986, Farming the Dust Bowl: a first-hand account from Kansas, University Press of Kansas, Lawrence, 255 pp.
Worster, D., 1979, Dust Bowl: the southern plains in the 1930s, Oxford University Press, New York, 277 pp.

California and Colorado River—Water Depletion
Hundley, N., 1975, Water and the West: the Colorado River compact and the politics of water in the American West, University of California Press, Berkeley, 395 pp.
Hundley, N., 1992, The great thirst: California and water, 1770s-1990s, University of California Press, Berkeley, 551 pp.
Miller, G.T., 1996, Living in the environment (ninth edition), Wadsworth Publishing, Belmont, 727 pp.
Potter, L.D. and C.L. Drake, 1989, Lake Powell, University of New Mexico Press, Albuquerque, 311 pp.

Chapter 3
Buffaloes and Marine Mammals—Animal Overkill
Ehrlich, P. and A. Ehrlich, 1981, Extinction: the causes and consequences of the disappearance of species, Random House, 305 pp.
Gard, W., 1959, The great buffalo hunt, Alfred A. Knopf, New York, 324 pp.
Kenyon, K.W., 1977, Caribbean monk seal extinct, Journal of Mammalogy, 58, 97-98.
King, J.E., 1956, The monk seal (genus Monachus), Bulletin of the British Museum (Natural History) Zoology, 3, 210-256.
Knudtson, P.M., 1977, The case of the missing monk seal, Natural History, 86, 8, 78-83.
Martin, P.S. and R.G. Klein, editors, 1984, Quaternary extinctions, University of Arizona Press, Tucson, 892 pp.
Roe, F.G., 1970, The North American buffalo: a critical study of the species in its wild state, University of Toronto Press, Toronto, 991 pp.
Ziswiler, V., 1967, Extinct and vanishing animals, Springer-Verlag, New York, 153 pp.

Passenger Pigeon and Dodo—Bird Overkill
Allen, D.L., 1968, Dove of no peace, Audubon, 70, 5, 55-57.
Blockstein, D.E. and H.B. Tordoff, 1985, Gone forever: a contemporary look at the passenger pigeon, American Birds, 39, 845-851.
Greenway, J.C., 1958, Extinct and vanishing birds of the world, American Committee for International Wold Life Protection, New York, 518 pp.

Halliday, T., 1978, Vanishing birds: their natural history and conservation, Holt, Rinehart and Winston, New York, 296 pp.
Schorger, A.W., 1955, The passenger pigeon, Outing Publishing, New York, 225 pp.
Ziswiler, V., 1967, Extinct and vanishing animals, Springer-Verlag, New York, 133 pp.

Gypsy Moth and Zebra Mussel—Introduced Species
Counts, C.L., 1986, The zoogeography and history of the invasion of the United States by Corbicula Fluminea (Bivalvia: Corbiculidae), American Malacological Bulletin, Special Edition, 2, 7-39.
Doane, C.C. and M.I. McManus, eds., 1981, The gypsy moth: research toward integrated pest management, United States Department of Agriculture, Forest Service, Technical Bulletin, 1584, 1-757.
Drake, J.A., H.A. Mooney, F. di Castro, R.H. Groves, F.J. Kruger, M. Rejmánek and M. Williamson, eds., 1989, Biological invasions, John Wiley, Chichester, 525 pp.
Klassen, W., 1989, Eradication of introduced arthropod pests: theory and historical practice, Entomological Society of America, Miscellaneous Publications, 73, 1-28.
Nalepa, T.F. and D. Schloesser, eds., 1993, Zebra mussels: biology, impacts and controla, Lewis Publishers, Boca Raton, 810 pp.
Page, J., 1990, Fly away, fly away, fly away home, Smithsonian, 21, 6, 76-83.
Silva, J.M., 1991, Gypsy moths: thwarting their ways, Agricultural Research, 39, 5, 4-11.
Windle, P.N., 1993, Harmful non-indigenous species in the United States, Office of Technology Assessment, United States Congress, Washington, 391 pp.

Chapter 4
Love Canal—Lois Gibbs
Brown, M.H., 1979, Laying waste: the poisoning of America by toxic chemicals, Pantheon Books, New York, 351 pp.
Deegan, J., 1987, Looking back at Love Canal, Environmental Science and Technology, 21, 328-331.
Dionne, E.J., 1982, U.S. finds Love Canal neighborhood is habitable, The New York Times, 15 July, A1 and B5.

Ember, L., 1989, Occidental agrees to store, treat Love Canal wastes, Chemical and Engineering News, 67, 25, 20-21.

Fowlkes, M.R. and P.Y. Miller, 1982, Love Canal: the social construction of a disaster, Federal Emergency Management Agency, Work Unit Report Number 6441E, Washington, 144 pp.

Gibbs, L.M., 1982, Love Canal: my story, State University of New York Press, Albany, 174 pp.

Kolata, G.B., 1980, Love Canal: false alarm caused by botched study, Science, 208, 1239-1242.

Levine, A.G., 1982, Love Canal: science, politics and people, D.C. Heath, Lexington, 263 pp.

Mauer, J., 1990, The availability of affirmative defenses of assumption of risk and the "sale defense" against common law public nuisance actions: United States v. Hooker Chemical and Plastics Corp., Natural Resources Journal, 30, 941-953.

Times Beach—Judy Piatt

Adler, S., 1988, Appelate court denies cover for cleanup costs, Business Insurance, 22, 10, 1 and 100.

Anderson, K., 1985, Living dangerously with toxic wastes, Time, 126, 15, 86-87.

Anonymous, 1990, Times Beach cleanup begins, Engineering News-Record, 225, 5, 37-38.

Brownlee, S., 1983, The dioxin dilemma, Discover, 44, 78-84.

Gallo, M.A., R.J. Scheuplein and K.A. Van der Heijden, 1991, Biological basis for risk assessment of dioxins and related compounds, Cold Spring Harbor Laboratory Press, Plainview, 501 pp.

Long, J., 1991, Court tells insurer to pay for dioxin cleanup, Chemical and Engineering News, 59, 38, 5-6.

Reinhold, R., 1983, Missouri dioxin cleanup: a decade of little action, The New York Times, 1 and 54.

Russakoff, D., 1983, U.S. offers to buy poisoned homes of Times Beach, The Washington Post, 23 February, A1-A2.

Steinberg, K.K., M.L. MacNeil, J.M. Karon, P.A. Stehr, J.W. Neese and L.L. Needham, 1985, Assessment of 2,3,7,8-TCDD tetrachlorodibenzo-p-dioxin exposure using a modified d-glucaric acid assay, Journal of Toxicology and Environmental Health, 16, 743-752.

Sun, M., 1983, Missouri's costly dioxin lesson, Science, 219, 367-369.

Cuyahoga River—Ben Stefanski

Anonymous, 1969, The cities: the price of optimism, Time, 1 August, 41.

Anonymous, 1978, The lagging cleanup of the Great Lakes, Business Week, 2536, 76-78.

Ellis, W.D., 1966, The Cuyahoga, Holt, Rinehart and Winston, New York, 302 pp.

Garlauskas, A.B., 1974, Water quality baseline assessment for Cleveland area: volume 1, synthesis, National Technical Information Service, Department of Commerce, Pb-238 353, 175 pp.

Loke, M., 1978, Urban parks getting aid in remedying some ills, The New York Times, 31 December, 21.

Seamonds, J.A., 1984, In Cleveland, clean waters give new breath of life, U.S. News and World Report, 96, 24, 68-69.

Centralia—Joan Girolami

Aurand, H.W., 1971, From the Molly Maguires to the United Mine Workers, Temple University Press, Philadelphia, 221 pp.

Broehl, W.G., 1964, The Molly Maguires, Harvard University Press, Cambridge, 200 pp.

De Kok, D., 1986, Unseen danger, University of Pennsylvania Press, Philadelphia, 299 pp.

Getlin, J., 1992, Mine fire still burns 30 years later, Valley News, 11 May, 7.

Jacobs, R., 1986, Slow burn: a photodocument of Centralia, Pennsylvania, University of Pennsylvania Press, Philadelphia, 152 pp.

Kroll-Smith, J.S. and S.R. Couch, 1990, The real disaster was above ground: a mine fire and social conflict, University Press of Kentucky, Lexington, 200 pp.

Lewis, A.H., 1964, Lament for the Molly Maguires, Harcourt, Brace and World, New York, 308 pp.

Logue, J.N., R.M. Stroman and K. Swarajah, 1991, The Centralia mine fire, Journal of Environmental Health, 54, 21-23.

Robbins, W., 1981, Pennsylvania town lives with fire that won't stop, The New York Times, 28 March, 1 and 10.

Chapter 5
London and Los Angeles—Harold Des Voeux
Meetham, A.R., D.W. Bottom, S. Cayton, A. Henderson-Sellers and D. Chambers, 1981, Atmospheric pollution: its history, origins and prevention (fourth edition), Pergamon Press, Oxford, 232 pp.

Stern, A.C., R.W. Boubel, D.B. Turner and D.L. Fox, 1984, Fundamentals of air pollution, Academic Press, New York, 530 pp.

Wagner, R.H., 1971, Environment and man, W.W. Norton, New York, 491 pp.

Wise, W., 1968, Killer smog, Rand McNally, New York, 181 pp.

World Health Organization, 1961, Air pollution, Columbia University Press, New York, 442 pp.

Lake Erie and Chesapeake Bay—Chesapeake Bay Foundation
Burns, N.M., 1985, Erie: the lake that survived, Rowman and Allanheld, Totowa, 320 pp.

Horton, T. and W.M. Eichbaum, 1991, Turning the tide: saving the Chesapeake Bay, Island Press, Washington, 328 pp.

Officer, C.B. and J.H. Ryther, 1980, The possible importance of silicon in marine eutrophication, Marine Ecology, 3, 83-91.

Officer, C.B., R.B. Biggs, J.L. Taft, L.E. Cronin, M.A. Tyler and W.R. Boynton, 1984, Chesapeake Bay anoxia: origin, development and significance, Science, 223, 22-27.

Ryther, J.H. and C.B. Officer, 1981, Impact of nutrient enrichment on water uses. In Neilson, B.J. and L.E. Cronin (editors), Estuaries and nutrients, Humana Press, Clifton, pp. 247-261.

Wagner, R.H., 1971, Environment and man, W.W. Norton, New York, 491 pp.

Acid Rain—Noye Johnson
Beamish, R.J., W.L. Lockhart, J.C. Van Loon and H.H. Harvey, 1975, Long term acidification of a lake and resulting effect on fishes, Ambio, 4, 98-102.

Firor, J., The changing atmosphere, Yale University Press, New Haven, 145 pp.

Fulkerson, W., R.R. Judkins and M.J. Sanghvi, 1990, Energy from fossil fuels, Scientific American, 263, 3, 128-135.

Johnson, N.M., R.C. Reynolds and G.E. Likens, 1972, Atmospheric sulfur: its effects on the chemical weathering of New England, Science, 177, 514-516.

Linzon, S.N., Effects of airborne sulfur pollutants on plants. In Nriagu, J.O. (editor), Sulfur in the environment. Part II: ecological impacts, John Wiley and Sons, New York, pp. 109-162.

Mohnen, V.A., 1988, The challenge of acid rain, Scientific American, 259, 2, 30-38.

Roberts, L., 1991, Learning from an acid rain program, Science, 251, 1302-1305.

Ogallala Aquifer—Groundwater Management Districts

Morgan, M.D., J.M. Moran and J.H. Wiersma, 1993, Environmental science, Wm. C. Brown Publishers, Dubuque, 480 pp.

Opie, J., 1993, Ogallala, water for a dry land, University of Nebraska Press, Lincoln, 412 pp.

White, S.E. and D.E. Kromm, 1995, Local groundwater management effectiveness in the Colorado and Kansas Ogallala region, Natural Resources Journal, 35, 275-307.

Zwingle, E., 1993, Wellspring of the high plains, National Geographic, 183, 3, 80-109.

Chapter 6

DDT—Rachel Carson

Brooks, P., 1972, The house of life: Rachel Carson at work, Houghton Mifflin, Boston, 350 pp.

Brooks, P., 1987, Courage of Rachel Carson, Audubon, 89, 1, 12-15.

Carson, R., 1962, Silent Spring, Houghton Mifflin, Boston, 368 pp.

Graham, F., 1970, Since Silent Spring, Houghton Mifflin, Boston, 333 pp.

Marco, G.J., R.M. Hollingsworth and W. Durham, 1987, Silent Spring revisited, American Chemical Society, Washington, 214 pp.

McCay, M.A., 1993, Rachel Carson, Twayne Publishers, New York, 122 pp.

Ozone Layer Depletion—Mario Molina and Sherwood Rowland

Bowman, K.P., 1988, Global trends in total ozone, Science, 239, 48-50.

Easterbrook, G., 1995, A moment on the Earth, Viking, New York, 745 pp.

Firor, J., 1990, The changing atmosphere, Yale University Press, New Haven, 145 pp.

Lemonick, M.D., 1992, The ozone vanishes, Time, 17 February issue, 60-68.

Molina, M.J. and F.S. Rowland, 1974, Stratospheric sink for chlorofluoromethanes: chlorine atomic-atalysed destruction of ozone, Nature, 249, 810-812.

Ray, D.L. and L. Guzzo, 1993, Environmental overkill, Harper Perennial, New York, 260 pp.

Schell, J., 1989, Our fragile Earth, Discover, October issue, 44-48.

Stolarski, R.S., 1988, The Antarctic ozone hole, Scientific American, 258, 1, 30-36.

Taubes, G., 1993, The ozone backlash, Science, 260, 1580-1583.

World Meteorological Organization, 1995, Scientific assessment of ozone depletion: 1944, World Meteorological Organization, Global Ozone Research and Monitoring Project, Report No. 37, 36 pp.

Global Warming, Part I—Roger Revelle

Abelson, P.H., 1990, Uncertainties about global warming, Science, 247, 1529.

Abelson, P.H., 1990, Global change, Science, 249, 1085.

Ahrens, C.D., 1988, Meteorology today, West Publishing, St. Paul, 582 pp.

Dumanoski, D., 1988, Steps to halt global warming must start now, analysts say, The Boston Globe, 18 July, 33-34.

Firor, J., 1990, The changing atmosphere, Yale University Press, New Haven, 145 pp.

Houghton, J., 1997, Global warming: the complete briefing, Cambridge University Press, Cambridge, 251 pp.

Idso, S.B., 1989, Carbon dioxide and global change: earth in transition, IBR Press, Tempe, 292 pp.

Jones, P.D. and T.M.L. Wigley, 1990, Global warming trends, Scientific American, 263, 2, 84-91.

Michaels, P.J., 1990, The greenhouse effect and global change: review and reappraisal, International Journal of Environmental Studies, 36, 55-71.

Morgan, M.D., J.M. Moran and J.H. Wiersma, 1993, Environmental science, Wm. C. Brown, Dubuque, 480 pp.

Officer, C. and J. Page, 1993, Tales of the Earth, Oxford University Press, New York, 226 pp.

Shabecoff, P., 1988, Major greenhouse impact is unavoidable, experts say, The New York Times, 19 July, C1 and C6.

White, R.M., 1990, The great climate debate, Scientific American, 263, 1, 36-43.

Global Warming, Part II—Wallace Broecker

Bradley, R.S., 1999, Paleoclimatology: reconstructing climates of the Quaternary, Academic Press, San Diego, 613 pp.

Broecker, W.S., 1997, Will our ride into the greenhouse future be a smooth one, GSA Today, 7, 5, 1-7.

Broecker, W.S., 1999, What if the conveyor were to shut down: reflections on a possible outcome of the great global experiment, GSA Today, 9, 1, 1-7.

Dansgaard, W., S.J. Johnsen, H.B. Clausen, D. Dahl-Jensen, N.S. Gundestrup, C.V. Hammer, C.S. Hvidberg, J.P. Steffensen, A.E. Sveinbjörnsdottir, J. Jouzel, and G. Bond, 1993, Evidence for general instability of past climate from a 250-kyr ice-core record, Nature, 364, 218-220.

Folland, C.K., T.N. Palmer, and D.E. Parker, 1986, Sahel rainfall and worldwide sea temperatures, 1901-85, Nature, 320, 602-607.

Imbrie, J. and K.S. Imbrie, 1979, Ice Ages: solving the mystery, Harvard University Press, Cambridge, 224 pp.

Jouzel, J., R.B. Alley, K.M. Cuffey, W. Dansgaard, P. Grootes, G. Hoffmann, S.J. Johnsen, R.D. Koster, D. Peel, C.A. Shuman, M. Stievenard, M. Stuiver, and J. White, 1997, Validity of the temperature reconstruction from water isotopes in ice cores, Journal of Geophysical Research, 102, 26471-26487.

Kerr, R.A., 1999, A new force in high-latitude climate, Science, 284, 241-242.

McPhee, M.G., T.P. Stanton, J.H. Morison, and D.G. Martinson, 1998, Freshening of the upper ocean in the Arctic: is perennial sea ice disappearing, Geophysical Research Letters, 25, 1729-1732.

Pomerance, R., 1999, Coral bleaching, coral mortality, and global climate change, http//www.stategov/www/gobal/global_issues/coral_reefs/990305_coralreef_rpt.html, 14 pp.

Stuart, A.J., 1991, Mammalian extinctions in the late Pleistocene of Northern Eurasia and North America, Biological Reviews, 66, 453-462.

Green Revolution—Norman Borlaug

Anonymous, 1999, U.S. plans new restrictions to rebuild fish populations, The New York Times, 28 April, A18.

Bourjaily, V., 1973, One of the Green Revolution boys, Atlantic Monthly, 231, 2, 66-70.

Crosson, P.R., 1992, Sustainable agriculture, Resources for the Future, 106, 14-17.

Davidson, M., 1988, A conversation with Norman Borlaug, USA Today, 117, 72-75.

Fogarty, M.J., A.A. Rosenberg and M.P. Sissenwine, 1992, Fisheries risk assessment, Environmental Science and Technology, 26, 440-446.

Hjort, H., 1914, Fluctuations in the great fisheries of Northern Europe, Conseil Permanent International pour l'Exploration de la Mer, Rapports et Procès-verbaux, 20, 1-228.

Miller, G.T., 1996, Living in the environment (ninth edition), Wadsworth Publishing, Belmont, 727 pp.

Morgan, M.D., J.M. Moran and J.H. Wiersma, 1993, Environmental science, Wm. C. Brown, Dubuque, 480 pp.

Pimentel, D., 1995, Agriculture, technology and natural resources. In Makofske, W.J. and E.F. Karlin, editors, Technology and global environmental issues, Harper Collins, New York, pp. 38-50.

PART II
Chapter 7

Bogue, D.J., 1984, Population, Encyclopedia Americana, 22, 402-408.

Masters, C.D., D.H. Root, and E.D. Attansi, 1990, World oil and gas resources—future production realities, Annual Reviews of Energy, 15, 23-51.

Chapter 8
Bogue, D.J., 1984, Population, Encyclopedia Americana, 22, 402-408.
Gerami, S., 1996, Women and fundamentalism: Islam and Christianity, Garland Publishing, New York, 178 pp.
Greeley, A.M., 1989, Religious changes in America, Harvard University Press, Cambridge, 137 pp.
Malthus, T.R., 1798, An essay on the principle of population as it affects the future improvement of society, J. Johnson, London, 396 pp.
Mernissi, F., 1991, Women and Islam: an historical and theological enquiry, Basil Blackwell, Oxford, 228 pp.
Miller, G.T., 1996, Living in the environment (ninth edition), Wadsworth Publishing, Belmont, 727 pp.
Noonan, J.T., 1986, Contraception: a history of its treatment by the Catholic theologians and canonists, Harvard University Press, Cambridge, 581 pp.
O'Brien, J.A., editor, 1968, Family planning in an exploding population, Hawthorn Books, New York, 222 pp.
Strobel, M., 1979, Muslim women in Mombassa, 1890-1975, Yale University Press, New Haven, 258 pp.
Winch, D., 1987, Malthus, Oxford University Press, Oxford, 117 pp.
World Resources Institute, 1992, World Resources, 1992-93, Oxford University Press, New York, 385 pp.
World Resources Institute, 1998, World Resources, 1998-99, Oxford University Press, New York, 369 pp.
Zwingle, E., 1998, Women and population, National Geographic, October, 36-55.

CHAPTER 9
Anonymous, 1999, Jane's fighting ships, Jane Information Group, Surrey, 912 pp.
Davis, G.R. et al., 1990, Energy for planet Earth, Scientific American, 263, 3, 55-165.
Knight, G., 1976, Concorde: the inside story, Stein and Day, New York, 176 pp.
Masters, C.D., D.H. Root and E.D. Attansi, 1990, World oil and gas resources—future production realities, Annual Reviews of Energy, 15, 23-51.

Munson, K., 1969, Aircraft of World War Two, Doubleday, New York, 256 pp.

Office of Management and Budget, 1998a, Budget of the United States government, fiscal year 1999: analytical perspectives, Office of Management and Budget, Washington, D.C., 586 pp.

Office of Management and Budget, 1998b, Budget of the United States government, fiscal year 1999: historical tables, Office of Mangement and Budget, Washington, D.C., 280 pp.

Office of Management and Budget, 1998c, Budget of the United States government, fiscal year 1999: appendix, Office of Management and Budget, Washington, D.C., 1227 pp.

Starr, C., 1971, Energy and power, Scientific American, 225, 3, 37-49.

Stone, R. and P. Szuromi, editors, 1999, Powering the next century, Science, 285, 677-711.

INDEX

Acid rain, 131-135, 196
Acid rain effects, fish, 133
Acid rain effects, trees, 133
Acidity, 131
Agricultural production, limits, 186-187
AIDJEX, 179-180
Amazonia, 21-29
Amsterdam Accord, U.N., 218-220
Amu Darya, 42-44
Anchovy, Peru, 191
Anoxia, 125
Apollo program, 226-227
Aral Sea, 41-47, 199-200
Arctic ice cover, 179-180
Arctic Oscillation, 180
Asian clam, 80
Augustine, Saint, 212-213

Babayev, A., 43
Biochemical oxygen demand, 122
Birth control, worldwide, 210-211
Black Northerns, 48
Black Sunday, 50-51
Boll weevil, 76-77
Borlaug, Norman, 184-186
Bormann, Herbert, 131
Brazil, 21-29-199-200
Broecker, Wallace, 172, 177
Buffaloes, decimation, 66-69

California Aqueducts, 60-62
Carbon cycle, global, 162-163
Carbon dioxide, atmosphere, 163-165
Carey, Hugh, 92
Caribbean monk seal, 69-70

Carson, Rachel, 141-151, 198
Carter, Jimmy, 94
Catholic church, 212-215
Center for Disease Control, 98-99
Center pivot irrigation, 137-138
Centralia mine, 102-112, 198
Cerrillo, Debbie, 89
CFCs, 153-154
Chesapeake Bay, 125-128, 130-131, 197
Chesapeake Bay Foundation, 130
Chestnut blight, 75-76
Christophe, Henry, 18-20
Clarke, Sir Andrew, 38
Cod, Georges Bank, 190
Coddington, John, 107
Cold war, 226-227, 235-236
Colorado River, 58-60
Columbus, Christopher, 17-18
Conness, John, 3
Cooke, Jay, 6
Coral, morality, 181-182
Crab wars, 127
Cuyahoga River, 99-102, 197-198

Darling, Jay Northwood, 2, 13-16
DDT, 147-151
Deforestation, 16-17
Des Voeux, Harold, 116-117
Dessalines, Jean-Jacques, 18-19
Dewling, Richard, 96-97
Dickens, Charles, 115
Dieldrin, 148-149
Dioxin, 98-99
Doane, Gustavus, 5
Dodo, extinction, 74-75
Domboski, Todd, 106
Dorigo, Vladimoro, 39-40
Doyle, Arthur Conan, 115-116

Dust Bowl, 47-56, 198
Dutch elm disease, 76
Duvalier, Francois, 20-21
Duvalier, Jean-Claude, 20-21

Energy, nuclear, 230
Energy, solar, 230
Energy, thermal, 230
Energy consumption, U.S., 223-224
Energy sources, alternative, 200-202, 221-236
Environmental Protection Agency, 99
Erie, Lake, 101-102, 128-130, 197
Eutrophication, cultural, 124
Evelyn, John, 114-115
Extinctions, mammals and birds, 65-66
Extinctions, Pleistocene mammals, 176-177

Family planning, 217-220
Federal budget, 1999, 231-233
Federal budget, 1999, research, 233-235
Female education, worldwide, 211-212
Fertility, United States, 207-208
Fish and Wildlife Refuges, 13-15
Fishing, limits, 187-192
Fly ash barrier, 106
Folland, C.K., 181-182
Fossil fuels, depletion, 201-202
Frasseto, Roberto, 36-38

Gasoline prices, 225-226
Gibbs, Lois, 88-94, 109-111
Girolami, Joan, 108-109
Global warming, historical record, 166-167
Global warming, model calculations, 167-169
Global warming, part I, 160-171

Global warming, part II, 171-182
Grain production, 184-186
Grain production, per capita, 186
Gray, James A., 38-39
Great Buffalo Hunt, 66-69
Green Revolution, 183-192
Greenhouse effect, 160-161
Greenland ice cores, 175
Groundwater Management Districts, 137-139
Gypsy moth, 77-79

Haddock, Georges Bank, 190
Haiti, 16-21, 199-200
Halocline, 124
Hayden, Ferdinand, 5
Hazardous chemicals, 91, 97
Heath, Clark, 96-97
Heptachlor, 148
Herring, Norway, 188-189
Hetch Hetchy Valley, 5, 11-12
High Plains, 47-56
Holocene climatic optimum, 179
Hooker Chemical and Plastics Corp., 84-86, 97
Hoover, Herbert, 58
Hopkins, Harry, 14
Hydrilla, 79
Hypoxia, 125

Ice Ages, 172-175
International Fund for Monuments, 38-39
Interstadials, 174-175
Irish potato famine, 205
Islamic faith, 215-216

Johnson, Noye, 131

Kara Kum Canal, 44
Kerr, Robert, 57-58
Killer smog, 118-119
Komet, 227-228

Langford, Nathaniel P., 5
Levintanus, Arkady, 46
Likens, Gene, 131
Limiting nutrient, 129-130
Little Ice Age, 179
Lo Grosso, Giuseppe, 33
Love Canal, 83-97, 198-199
Love Canal Homeowners Association, 89-97

MacAndrew, John, 38
Maccoun, Robert, 38
Maheu, René, 34-35
Malthus, Thomas Robert, 203-205
Malthus Dismal Theorem, 204-205
Mammoth Hot Springs, 6
Marghera, Porto, 32-33
Martinson, Douglas, 180
McPhee, Miles, 180
Medfly, 77
Medieval Warm Period, 179
Mendes, Chico, 27-28
Mestre, 32-33
Milankovitch cycles, 173
Mine fire, 102-112
Molina, Mario, 158
Molly Maguires, 104-105
Moral and religious leadership, 212-216
Moran, Thomas, 6
Morison, James, 180
Muiniak, 45
Muir, John, 2, 8-10
Mulholland, William, 161
Murphy, Sister Honor, 111-112

NAPAP, 133-135
National Forests, 12-13
National Parks, 2-8
National Smoke Abatement Society, 116-117
National Wildlife Federation, 15

Norwich, Lord, 38
Natural resources depletion, 221-225
Nutrients, 123, 129-130

Ocean conveyor belt, 177-182
Ogallala Aquifer, 56, 135-139, 197
Oil reserves, 201-202, 224
Oil shales, 224-225
Old Faithful Geyser, 6
Orellana, Francisco de, 22-24
Ozone layer, 151-160, 196
Ozone layer, Antarctic, 155
Ozone layer, formation and depletion, 152-154

Palmer, T.N., 181-182
Parker, D.E., 181-182
Passenger pigeon, extinction, 70-74
Pesticides, 141-151
Petion, Alexandre, 19
Photochemical haze, Los Angeles, 119-122, 196
Photosynthesis, 123
Piatt, Judy, 97-98
Picciano, Dante, 93-95
Pinchot, Gifford, 2, 12-13
Population, decimation, 208-210
Population growth, 200-202, 203-220
Population growth, history, 205-210
Precautionary principle, 171, 182

Radiocarbon ages, 177
Raymond, Israel Ward, 3
Research and development, 226-229
Revelle, Roger, 163-164
Reynolds, Robert, 131
Roosevelt, Franklin, 13-15
Roosevelt, Theodore, 8, 12-13
Rowland, Sherwood, 158
Rubber, Amazon, 25-26

Sahel, drought, 181-182
Scrubbers, 135
Sea Around Us, The, 143-145
Sea lamprey, 79-80
SHEBA, 180
Sierra Club, 11-12
Silent Spring, 146-151, 198
Smog, London, 113-119, 196
Species, introduced, 75-81
Species decimation, 63-75
SST, 228-229
Stadials, 174-175
Stanton, Timothy, 180
Starling, 80-81
Stefanski, Ben, 101-102
Svobida, Lawrence, 51-54
Swordfish, 228-229
Syr Darya, 42-44

Taliban, 216
Tar sands, 224-225
Temperature inversion, 126-127
Thermocline, 124
Thomas, Lewis, 95-96
Times Beach, 97-99, 199
Tonton Macoute, 20
Toussaint L'Ouverture, 18
Trash fish, 128
Tropical rain forests, 26-29

Ultraviolet radiation, 151-160

Venice, 31-41, 199-200
Venice in Peril, 38
Voghera, Giorgio, 38
Volpe, Count Guiseppe, 32
Volpe, Countess Cicogna, 39
Washburn, Henry, 5
Whalen, Robert, 90-92
White-Stevens, Robert, 150
Worster, Donald, 54-56

Yellowstone National Park, 5-7

Yosemite National Park, 3-5
Younger Dryas, 174-177

Zebra mussel, 80,